Frontiers in Clinical Drug Research:
Diabetes & Obesity
(Volume 2)

Editor

Atta-ur-Rahman, *FRS*

Kings College
University of Cambridge
Cambridge
UK

CONTENTS

PREFACE

Frontiers in Clinical Drug Research – Diabetes and Obesity Volume comprises seven comprehensive chapters discussing novel approaches to combat diabetes and obesity.

In Chapter 1, Kaji discusses the clinical trials conducted for evaluating the current, emerging and future medications for treating Type 2 diabetes and obesity. They also highlight the new drug candidates that are under consideration.

Overweight and obesity are the risk factors for type 2 diabetes and they also influence the overall prognosis. In Chapter 2, Abel and Lengyel emphasize that in order to prevent the development of complications, the antidiabetic therapies are designed to reduce and maintain the glucose concentrations.

Kidney plays a vital role in maintaining the level of glucose production and absorption. The major portion of glucose that enters into the kidney is reabsorbed in the tubular system, which helps in maintaining the glucose plasma level. Due to certain malfunctioning, glucose may not get properly reabsorbed in the tubular system that can lead to the condition of glycosuria. Sodium-coupled glucose transporters 2 (SGLTs-2) play an important role in glucose reabsorption. In Chapter 3 Papazafiropoulou and Kardara highlight the role of Sodium-Glucose Co-Transport Inhibitors as a new therapeutic approach for type 2 diabetes mellitus.

Kanazawa draws attention towards the impact of Osteocalcin on glucose metabolism in Chapter 4. Osteocalcin prevents the risk of diabetes mellitus by increasing the expression of insulin in pancreatic β-cells as well as adiponectin in adipocytes. Kanazawa presents an overview of the clinical studies, suggesting that glucose homeostasis and bone metabolism is linked together through the action of osteocalcin.

In Chapter 5, Channabasappa and Prasanna Kumar review the various new targets for diabetes and they propose that long acting GLP-1 (Glucagon-like Peptide-1) analogues, DPP-4 (Dipeptidyl peptidase-4) inhibitors and newer gliptins may be important in the discovery of anti-diabetic therapies. Similarly, in Chapter 6, Li *et al.* focus on incretin impairment that commonly exists in both obesity and diabetes, so it is necessary to work on incretin based therapies for the prevention and cure of diabetes and obesity.

In Chapter 7, Vallianou and colleagues discuss the importance of GLP-1 receptor agonists, and the 5-HT2C selective serotonin receptor agonists oxyntomodulin, oxytocin, orexin and PPARγ (peroxisome proliferator agonist receptor γ) that seem to be effective in the eradication of obesity.

I am very grateful to all the authors for their outstanding contributions. I would also like to appreciate the efforts of the dedicated team of Bentham Science Publishers, especially Dr. Faryal Sami (Assistant Manager Publications), Mr. Shehzad Naqvi (Senior Manager Publications) and Mr. Mahmood Alam (Director Publications).

Prof. Atta-ur-Rahman, FRS
Kings College
University of Cambridge
Cambridge
UK

CONTRIBUTORS

Angelos A. Evangelopoulos	Roche Hellas Diagnostics Athens, Greece
Athanasia K. Papazafiropoulou	1st Department of Internal Medicine & Diabetes Center General Hospital of Piraeus "Tzaneio", Greece
Christos E. Kazazis	Leicester University, London, UK
Gabriella Lengyel	2nd Department of Internal Medicine, Medical Faculty, Semmelweis University, Budapest, Hungary
Hidesuke Kaji	Division of Physiology and Metabolism, University of Hyogo, Akashi, Hyogo, Japan
Ippei Kanazawa	Department of Internal Medicine 1, Shimane University Faculty of Medicine, Shimane, Japan
Krishnan M. Prasanna Kumar	Center for Diabetes and Endocrine Care, Banglore, India
Marina S. Kardara	Health Center of Erymantheia, Achaia, Greece
Minglong Li	Department of Endocrinology, Provincial Hospital affiliated to Shandong University, Jinan, China
Natalia. G Vallianou	Evangelismos General Hospital Athens, Greece
Shivaprasad Channabasappa	Department of Endocrinology and Metabolism, Vydehi Institute of Medical Sciences & Research Centre, Banglore, India
Tatjana Ábel	Outpatient Department, Military Hospital, Budapest Hungary; Faculty of Health Sciences, Semmelweis University, Budapest, Hungary
Xianglan Sun	Department of Endocrinology, Provincial Hospital affiliated to Shandong University, Jinan, China
Yao Wang	Shandong Institute of Endocrine and Metabolic Diseases, Shandong Academy of Medical Sciences, Jinan, 250062 Shandong, China

Frontiers in Clinical Drug Research:
Diabetes & Obesity
(Volume 2)

2

Frontiers in Clinical Drug Research – Diabetes and Obesity

Volume # 2

Editor: Prof. Atta-ur-Rahman

ISSN (Online): 2352-3220

ISSN: Print: 2467-9607

eISBN: 978-1-68108-185-4

ISBN: 978-1-68108-186-1

© 2016, Bentham eBooks imprint.

Published by Bentham Science Publishers –
Sharjah, UAE. All Rights Reserved.

BENTHAM SCIENCE Bentham Books

CHAPTER 1

The Current, Emerging and Future Medications for Type 2 Diabetes and Obesity

Hidesuke Kaji[*]

Division of Physiology and Metabolism, University of Hyogo, Akashi, Hyogo, Japan

Abstract: This chapter firstly presents selected clinical studies to evaluate the current drugs as well as emerging new drugs for type 2 diabetes mellitus (T2DM) and obesity, and secondly presents future drug candidates under pharmacological research. The current drugs are α-glucosidase inhibitors, glinide, sulfonylurea (SU), incretin-based therapies, including dipeptidyl peptidase 4 (DPP-4) inhibitors and glucagon-like peptide-1 (GLP-1) receptor agonist, thiazolidinediones (TZDs), biguanides, insulin, in particular, short acting and long acting insulin analogs. Anti-obesity drugs available at present are intestinal lipase inhibitor. Most of central regulators of appetite are not used mainly because of their cardiovascular adverse events. Better glycemic control was compared among various drug combinations; DPP-4 inhibitors *vs.* TZDs add-on to metformin, DPP-4 inhibitors *vs.* GLP-1 receptor agonist add-on to metformin or other oral hypoglycemic therapy (LEAD-6), DPP-4 inhibitors *vs.* placebo with and without SU, short acting insulin analog *vs.* DPP-4 inhibitors add-on to metformin (EASIE study) or *vs.* placebo on GLP-1 receptor agonist add-on to metformin and SU. Effects on morbidity and mortality have been evaluated between intensive and conventional glycemic control. Intensive glycemic control caused more frequent severe hypoglycemia and did not necessarily ameliorate prognosis being much better as compared with conventional glycemic control. Effects were also evaluated among the following drugs; DPP-4 inhibitors *vs.* SU, glinide (NAVIGATOR study), basal supported oral therapy (BOT), basal plus or basal bolus (ORIGIN trial), and insulin *vs.* other anti-hyperglycemic therapies. Emerging new drugs for T2DM and obesity are salt glucose cotransporter 2 (SGLT2) inhibitors, glucokinase activators, and anti-obesity drugs such as lorcaserin and phentermine-topiramate, both of which are central appetite regulators with less adverse events. These are prospective drugs but still need investigations to decide their clinical usefulness and safety. Pharmacological research for T2DM and obesity are underway. Various new drugs such as the following may be promising; cyclin-dependent kinase 5 inhibitor, leptin, adiponectin mimetics, SIRT1 activators, neuropeptide Y2 receptor antagonist, 11β-hydroxysteroid dehydrogenase type 1 inhibitor, interleukin-6, betatrophin, chemerin, dilauroyl phosphatidylcholine, 1- deoxynojirimycin, fibroblast growth factor(FGF)21, FGF1 and stearoyl-coenzyme A desaturase-1 inhibitors.

---

*Corresponding author Hidesuke Kaji: Division of Physiology and Metabolism, University of Hyogo, Akashi, Hyogo, Japan; Tel: 81-78-925-921; E-mail: hidesuke_kaji@cnas.u-hyogo.ac.jp

Keywords: Type 2 diabetes mellitus, obesity, drug, randomized control trial, pharmacological research, metformin, incretin-based therapies, DPP-4 inhibitors, GLP- 1 receptor agonist, insulin, glycemic control, morbidity, mortality

INTRODUCTION

Type 2 diabetes mellitus (T2DM) and obesity are lifestyle-related diseases especially common in the developed countries. These are serious risks for cardiovascular disease and shortening of life span. Diet and exercise are most important for their primary prevention and treatment. However, they are not easy to continue and sometimes require medications as an additional treatment. Various drugs are available now but are not yet perfect because of unsatisfactory efficacy or utility, adverse events, and cost issues. Better prognosis can be achieved by good control of blood glucose levels as well as body composition and by prevention of micro-vascular and macro-vascular complications. At first, this section deals with the efficacy and safety of current drugs recently evidenced by clinical studies and secondly those emerging new drugs which are still under ongoing clinical trials. The third section reviews the future drugs under pharmacological research for clinical applications to T2DM and obesity.

1. THE CURRENT MEDICATIONS FOR T2DM AND OBESITY (Fig. 1)

Mono-Therapy by the Current Drug

1) α-Glucosidase Inhibitors (α-GI)

α-GI (acarbose, voglibose, and miglitol) are pseudo-carbohydrates that competitively inhibit α-glucosidase expressed in the intestinal cells that hydrolyze non-absorbable oligosaccharides and polysaccharides into absorbable mono-saccharides. This slows absorption of carbohydrates and stops the postprandial rise in plasma glucose. α-GI can be used as a first choice in newly diagnosed T2DM when diet and exercise treatment or mono-therapy with oral drugs for T2DM and insulin are insufficient to control diabetes. α-GIs are contraindicated in subjects with diabetic ketoacidosis or Crohn's disease, ulcerative colitis, sub-ileus or in patients being to ileus. Since α-GIs do not degrade carbohydrates into glucose, some undigested carbohydrates are likely to be delivered to the colon. The complex carbohydrates are digested by bacteria in the colon with resultant adverse events such as flatulence (78 % of patients) and diarrhea (14 % of patients). Since these events occur dose-dependently, it is recommended to prescribe with a low dose and gradually increase the desired dose. A few cases of hepatitis have been reported with acarbose use, which regressed when the drug was stopped [1]; therefore, hepatic function should be monitored before and during this drug use.

2) Glinide

Glinide drugs (nateglinide, mitiglinide, repaglinide) are orally available, non-sulfonylurea (SU), insulin secretagogues. These drugs stimulate insulin secretion by blocking ATP-sensitive potassium (K_{ATP}) channels (Kir6.2/SU receptor 1 [SUR1]) in pancreatic β cells through binding to SUR1 with the binding site not identical to that of SUs. These agents are relatively short-acting, suitable for the treatment of postprandial hyperglycemia. The D-phenylalanine derivative, nateglinide, is a drug prescribed just before meals to enhance early insulin secretion and then reduce prandial plasma glucose concentrations. Because its effects are glucose-dependent [2] and are rapidly reversed [3], nateglinide can regulate hyperglycemia after meals with minimal postprandial hyperinsulinemia [4] and a low frequency of hypoglycemia [5, 6]. Repaglinide is a non-SUs benzoic acid derivative [7-9]. Repaglinide has a functional high-affinity binding site, which differs from that of SUs and nateglinide [10]. In *in vitro* study from isolated rat islets [2], the action of repaglinide failed to demonstrate glucose concentration-dependent sensitization unlike a glucose-dependent insulinotropic effect by nateglinide. Mitiglinide selectively acts on the Kir6.2 of pancreatic ß-cells and has higher affinity to the channel than other glinide drugs, repaglinide and nateglinide [11]. Mitiglinide also ameliorates hyperglycemia after meals in subjects with T2DM through both an insulin-mediated indirect and direct effect on hepatic glucose metabolism [12].

3) Sulfonylurea (SU)

SUs (glyburide/glibenclimide, glimepiride, glipizide, gliclazide) are insulin secretagogues used in T2DM for long time, which are inexpensive and effective. SUs stimulate insulin release by binding to the complex of SUR1 and K_{ATP}, causing membrane depolarlization, subsequent calcium channel opening to raise cytosolic calcium, and then stimulating insulin exocytosis. Since stimulation of insulin release by SUs is independent of ambient glucose levels, hypoglycemia frequently occurs in particular by glyburide [13]. Weight gain of 1-3 kg is also common. Other side effects are not common.

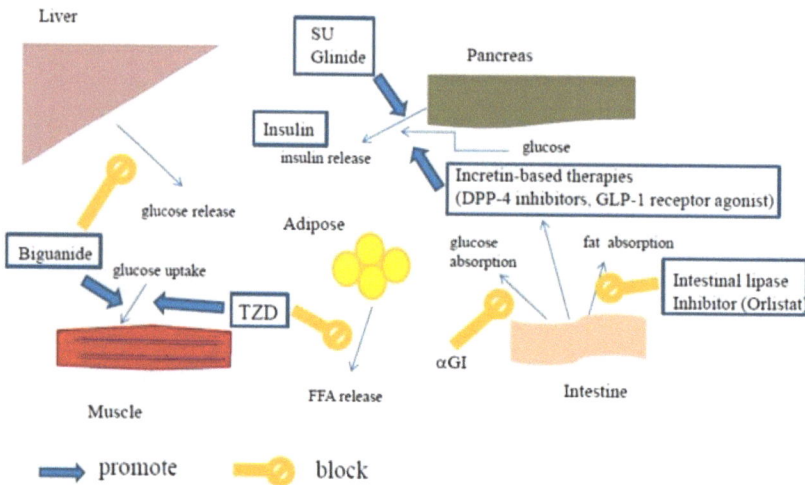

Figure 1: Major target organs and actions of existing drugs in T2DM and obesity. SU: sulfonyl urea, TZD: thiazolidinedione, DPP-4: dipeptidyl peptidase 4, GLP-1: glucagon like peptide-1 α-GI: α-glucosidase inhibitor, FFA: free fatty acid

4) Incretin-based Therapies

Incretins, intestinal hormones released in response to the ingestion of food, potentiate the insulin response to glucose. Human incretin is composed of mainly 2 peptide hormones, gastric inhibitory polypeptide (GIP) and glucagon like peptide-1 (GLP-1). GIP is released from K cells on the proximal intestine and GLP-1 is mainly secreted from the L cells on the distal intestine. Plasma concentrations of GIP and GLP-1 are very low in the fasting state, and rapidly increase after food intake.

Both incretins enhance insulin release and stimulate pancreatic ß-cell growth. GLP-1 also binds to α-cells to inhibit glucagon secretion along with other receptors in a variety of tissues to slow down gastric emptying, and promote satiety to decrease food intake [14-18]. The following two types are now available incretin-based therapies.

Dipeptidyl Peptidase 4 (DPP-4) Inhibitors

Half-lives of endogenous GLP-1 and GIP are very short, 2 minutes and 5-7 minutes, respectively, because they are quickly degraded by the ubiquitous enzyme, DPP-4 [14]. DPP-4 is widely expressed in a variety of tissues, including the brain, kidney, lung, adrenal gland, liver, intestine, spleen, testis, pancreas, as well as on the surfaces of lymphocytes and macrophages [14, 15]. Since DPP-4 degrades, biologically active GLP-1 is only 10-20 % of total plasma GLP-1 [18]. By the enzymatic inhibition of DPP-4, plasma concentrations of active GIP and GLP-1 can be increased 2- to 3-fold [19-21]. The DPP-4 inhibitors (sitagliptin, saxagliptin, vildagliptin, alogliptin, linagliptin, tenegliptin, and anagliptin) currently available are oral agents of small molecules that provide nearly complete and long-lasting inhibition of DPP-4. Most DPP-4 inhibitors are excreted by kidney, while linagliptin is excreted by bile duct. In clinical trials, DPP-4 inhibitors reduce HbA1c by 0. 7- 0. 9 % when used as mono-therapy or when added to metformin or thiazolidinediones without compromising tolerability [22]. Randomized, double-blind, placebo-controlled trial in forty one subjects demonstrated that insulin secretion was enhanced in patients with T2DM treated by vildagliptin. Vildagliptin caused small increases in fasting plasma GLP-1 in patients treated with metformin but not in those with diet therapy alone. The β cell effects of vildagliptin were observed in almost all treated subjects regardless of GLP-1 concentration, suggesting that vildagliptin can decrease fasting, as well as postprandial, glucose levels in T2DM subjects. Moreover, they suggest that there are additional mechanisms underlying action of DPP-4 inhibitors beyond potentiating incretin effects after meals [23]. However, these drugs include a warning about risk of pancreatitis and this should be factored into decision making.

GLP-1 Receptor Agonist

GLP-1 receptor agonists (exenatide/ lixisenatide, liraglutide) are synthetic peptide that mimic GLP-1.

Exenatide, a GLP-1 receptor agonist, ameliorates glycemic control in T2DM subjects by various mechanisms of action: increased glucose-dependent insulin secretion, reduced glucagon secretion after meals, slowed gastric emptying, and enhanced satiety [24, 25]. Exenatide is an exendin-based GLP-1 receptor agonist and is 53 % identical to an amino acid sequence of human GLP-1. Exenatide has 2- 3 hours of plasma half-life after subcutaneous injection. A long-acting exenatide (exenatide QW) recently developed is effective when injected once per week. Weekly administration of 2 mg exenatide QW results in therapeutic range within 2 weeks and steady-state 6-7 weeks after start of treatment [26, 27].

Lixisenatide (AVE0010) is a new selective once-daily GLP-1 receptor agonist in development for the treatment of T2DM [28]. It is a 44-amino acid peptide with the amidation of C-terminal end and shares structure with exendin-4 (the main difference is the addition of 6 lysine residues at the C terminal end) [29]. Lixisenatide is highly selective for the GLP-1 receptor having about fourfold higher affinity for the GLP-1 receptor than native human GLP-1 [29]. Liraglutide is GLP-1 receptor agonist whose molecule is arginine at position 34 of GLP-1 instead of lysine and C-16 palmitic acid side chain attached *via* a glutamyl spacer to lysine at position 26 of GLP-1. Liraglutide has plasma half-life of 12 hours after subcutaneous injection. The GLP-1 receptor agonists reduce plasma HbA1c by 1-1.5 %, with liraglutide 20-30 % more potent than exenatide [30]. Both drugs cause weight loss of ~2.5-4 kg over 6 months [31, 32]. The adverse events of GLP-1 receptor agonists are mainly nausea and vomiting. Association of pancreatitis with GLP-1 receptor agonists has been reported as DPP-4 inhibitors, although the mechanism to explain this association remains unknown and animal studies are conflicting on this question [32-34].

5) Thiazolidinediones (TZDs)

TZDs are agonists of peroxisome proliferator-activated receptor-γ (PPAR- γ) and play the pivotal role in amelioration of insulin resistance and the circulatory system. PPARs are a subfamily of nuclear receptors that are similar to the thyroid hormone as well as retinoid receptors [35]. PPARs are transcription factors that are activated by ligand and regulate target gene transcription. PPARs regulate the transcription of many genes important for the pathway of glucose and lipid metabolism and play a pivotal role in adipocyte differentiation [36, 37]. Three PPARs have been identified —PPARα, PPAR-β (or δ) and PPAR-γ. PPAR-γ is the most abundant in adipose tissue, but also in pancreatic β-cells, vascular endothelium, macrophages, and skeletal muscle [38-40]. It is still unclarified whether TZDs-induced improvement of insulin resistance is due to the direct effects on skeletal muscle and liver, indirect effects mediated by adipocytokines, like adiponectin, or some combination of these. TZDs have been available for the control of hyperglycemia in T2DM subjects since 1997. Troglitazone was the first clinically available TZDs, but was prohibited to use because of liver toxicity causing fulminant hepatitis. Currently, only pioglitazone and rosiglitazone are available for patients with T2DM. TZDs cause average reduction in HbA1c of 0.5-1.5 % and are effective as mono-therapy and as additive therapy to metformin, SUs, or insulin. Pioglitazone reduces plasma triglyceride (TG) by 10-15 %, and raises high-density lipoprotein (HDL) cholesterol levels, probably attributable to a dual effect on PPARα and PPARγ. TZDs have been used in the treatment of non-alcoholic steatohepatitis (NASH). The meta-analysis showed that TZDs were significantly more efficient than placebo in ameliorating ballooning degeneration, lobular inflammation and steatosis with combined odds ratios (ORs) of 2.11 (95 % confidence intervals [CI], 1.33-3.36), 2.58 (95 % CI, 1.68-3.97) and 3.39 (95 % CI, 2.19-5.25), respectively. When pioglitazone was analyzed alone, the amelioration in fibrosis with pioglitazone *vs.* placebo (combined OR 1.68 [95 % CI, 1.02-2.77]) was statistically significant [41]. The most common adverse events of the TZDs are weight gain of 2-4 kg over the first year of treatment and edema up to 10 % of patients. The adverse event of concern is congestive heart failure due to plasma volume expansion [42]. It has been recently reported that rosiglitazone increases the risk of cardiovascular events such as myocardial infarction and stroke. Therefore, TZDs are not recommended to use in T2DM subjects with moderate to severe heart failure [43]. The other adverse events are the increased risk of bone fracture [44, 45]. TZDs, especially pioglitazone, are associated with an increased risk of bladder cancer among T2DM adults based upon the hypothesis by the limited evidence [46].

6) Biguanide

How does biguanide (metformin) improve hyperglycemia remains to be clarified. In diabetic subjects, metformin decreases hepatic glucose output as a major effect, primarily by the inhibition of gluconeogenesis, but by the stimulation of glucose uptake by skeletal muscles as a lesser effect [47]. It has been reported that metformin activates liver and skeletal muscle adenosine monophosphate-activated protein kinase (AMPK), an enzyme normally activated by adenosine monophosphate, a cellular signal for increased energy needs [48]. Metformin lowered HbA1c levels by about 1.0-1.5 % [49 -51]. Metformin mono-therapy is as effective as SU mono-therapy [51, 52]. Metformin appears to have beneficial effects beyond glycemic control such as weight loss, or at least no weight gain. Improvements have also been noted in lipid profile such as reductions in plasma levels of FFA [53], TG and very low-density lipoproteins [54] in T2DM subjects with dyslipidemia. Increased levels of plasminogen activator inhibitor-1 [55] and C-reactive protein [56], both of which are cardiovascular risk markers, were also decreased by metformin. Metformin is approved for use in diabetes either as mono-therapy or in combination with other oral hypoglycemic agents and insulin. It is recommended as first choice therapy for obese subjects with T2DM [57]. Metformin is relatively not expensive. Gastrointestinal adverse events including abdominal discomfort, anorexia, bloating and diarrhea are dose-dependently observed in 10-15 % of subjects. The reason for these events remains unknown, but, like αGI, metformin has been associated with decreased intestinal glucose absorption [58]. These adverse events usually improve during use and are minimal if started at a low dose and slowly titrated upward. Discontinuation of treatment caused by adverse events occurs in less than 4 % of subjects [59]. Since insulin release is unchanged, hypoglycemia is not an adverse event of metformin when used as mono-therapy. Similarly, unlike some of the other drugs, weight gain is not an adverse events, and weight loss occurs in some patients [59]. Although lactic acidosis was frequent adverse events with the currently unavailable biguanide phenformin, it was rare adverse events with still available biguanide metformin as long as hepatic and renal function was not impaired. In a recent Cochrane database systematic review, an incidence of lactic acidosis was 8.4 cases per 100,000 patient-years in the metformin treated T2DM subjects and 9 cases per 100,000 patient-years in the other glucose-lowering drug treated group [60]. Metformin is contraindicated in patients with moderate to severe renal, hepatic or cardiac dysfunction which are risk factors for lactic acidosis or drug accumulation.

7) Insulin

Exogenous insulin is one of the most established blood glucose-lowering therapies, and its use in patients with T2DM has grown markedly over recent years [61], consequent upon findings from UKPDS and the availability of analog insulin that have increased comfort with insulin initiation and titration. Early insulin therapy has recently been recommended in guidelines from the American Diabetes Association and European Association for the Study of Diabetes [62], but the risk-benefit profile of exogenous insulin in the management of patients with T2DM has also undergone scrutiny [63 -66]. Recent retrospective cohort study using data from the UK General Practice Research Database, 2000-2010, has shown that insulin treatment was associated with an increased risk of diabetes-related complications, cancer, and all-cause mortality in patients with T2DM [67].

Short Acting Insulin Analogs

These analogs all have an acute effect (within 30-60 minutes) with a peak action within 2 hours, allowing for control of postprandial glucose elevations when injected within 5 minutes before meals [68]. Insulin lispro is a peptide in that the amino acids proline at positions 28 is inverted by lysine at positions 29 [69-71]. As it is unstable, zinc is added to the lispro formulation in order to promote self-association into hexamers, which increases stability but is more rapidly absorbed compared with regular human insulin. Insulin aspart is a peptide in that the amino acid residue at position 28 on the B-chain of the human insulin molecule is substituted with aspartic acid, thereby causing charge repulsion to inhibit the formation of hexamers [72].

Insulin glulisine is a peptide in that asparagine is replaced by lysine at position 3 and that lysine is replaced by glutamic acid at position 29 on the B-chain of the human insulin [73]. These substitutions render mono- and dimeric glulisine molecules stable in the drug formulation solution, thereby causing more quick absorption than other insulin analogs [72, 74]. This difference in the structure explains the more rapid absorption and action of glulisine when compared with lispro observed in lean-to-obese subjects [73, 75].

Long Acting Insulin Analogs

Insulin glargine is a long acting insulin analog in that a C -terminal of the B-chain is elongated by two arginines and asparagine of the A-chain is replaced by glycine in position 21. From these modifications the isoelectric point is shifted from a pH of 5.4 to 6.7. This shift makes glargine less soluble at physiological pH compared with the regular insulin. Consequently, glargine has a longer duration of action lasting approximately 24 hours, without apparent peak in activity [76]. In add-on lantus® to oral hypoglycemic agents (ALOHA) sub-analysis, a simplified and pragmatic dose calculation formula for T2DM patients starting glargine basal supported oral therapy (BOT) optimal daily dose at 24 weeks= starting dose (0. 15 x weight) + incremental dose (baseline HbA1c - target HbA1c + 2) [77]. Insulin detemir is a neutral, soluble, long-acting insulin analog in which threonine is deleted from position at 30 on the B-chain of the human insulin and the ε-amino group of lysine B29 is acetylated with a 14-carbon myristoyl fatty acid [78, 79]. By this fatty acid modification, detemir can reversibly bind to the albumin, thereby causing the prolongation of this analog action [80]. Detemir can remain in a liquid after subcutaneous injection because of its solubility at neutral pH. This property contrasts with NPH insulin, which is a preformed crystalline precipitate suspension, and glargine, which is an acidic solution that precipitates in the subcutaneous tissue after injection. The low variability of detemir in pharmacokinetic and pharmacodynamic properties may be attributable to its solubility [78, 79]. Insulin degludec is a new-generation basal insulin, designed to form soluble multi-hexamers on subcutaneous injection, resulting in an ultra-long action profile [81] and reduced within-individual variability [82]. Degludec differs from insulin in that the amino acid threonine at position 30 on the B-chain of the human insulin was deleted and lysine at B29 was acylated by hexadecandioyl using glutamic acid as spacer connecting them. This structure easily makes insulin dihexamer, causing delayed absorbtion. Degludec was non-inferior to glargine in the HbA1c level. The meta-analysis showed that fasting plasma glucose and overall and hypoglycemia at night were significantly lower in number with degludec compared with glargine. The overall physical health component score at endpoint was significantly better with degludec compared with glargine [+0.66 (95 % CI, 0.04-1.28)], mainly because of a difference of the pain score [+1.10 (95 % CI, 0.22-1.98)]. In the mental domains, vitality was significantly higher with degludec *vs.* glargine [+0.81 (95 % CI, 0.01-1.59)] [83].

8) Anti-Obesity Drugs

Obesity is defined as the excessive accumulation of body fat. When increased risk of serious illness is associated, obesity is a disease that should be treated to lower body weight. In clinical practice, body mass index (BMI) and waist circumference are good demographic indicators of obesity and visceral obesity, respectively. Treatment options are dependent on these indicators, and adverse health consequences at present or in the future. T2DM and obesity are mutually dependent as treatment of each disease affects the other. In addition to diet and exercise, drug can be used in adults with a BMI of at least 30 kg/m^2 or of at least 27 kg/m^2 if they have an obesity-related illness [84].

Intestinal Lipase Inhibitor (Orlistat)

Orlistat is a lipase inhibitor in stomach and pancreas, leading to a decrease in lipolysis and absorption of dietary fat [85]. Orlistat is the only drug proved to be effective. Orlistat has been effective to lower body weight and long term management of obesity in two large multi-center randomized double-blind trials for 2 years, with 1187 [86] and 743 obese patients [87]. Patients given orlistat had siginificant weight loss compared to those given placebo (10.3 kg *vs.* 6.1 kg) in year 1. Patients continuously given orlistat in year 2 regained body weight half of those switched to placebo and those continuously given placebo regained weight of 2.5 kg whereas those switched to orlistat lost weight of 0.9 kg [86]. Similar results were shown in the other report [87]. When patients were given orlistat, plasma insulin levels at the end were lower compared to the levels at the start of the study, whereas these levels were not significantly different when given placebo (64.5 *vs.* 84.0 pmol/L and 86.4 *vs.* 86.3 pmol/L, respectively, p = 0.04 between treatment groups). In the same study, increase in fasting serum glucose levels were lower in the orlistat group than the placebo group (+0.06 mmol/L *vs.* +0.26 mmol/L, respectively, p = 0.001 between treatment groups). Moreover, orlistat also caused a greater improvement in serum insulin and glucose levels after 1 and 2 years compared to the placebo group at the end of the multicenter European study [86]. Orlistat use is reportedly associated with minor gastrointestinal adverse events but also with a risk increase in severe hepatic events. On the other hand, a meta-analysis of clinical trial data including 10000 patients showed no evidence that orlistat was associated with impaired liver function [88] and a likely mechanism of action has not been identified. By using the self-controlled case series design, Douglas *et al.* were able to establish that the increased risk of liver injury did not change between just before and after starting treatment, strongly suggesting that the association is not a causal relation [89].

Central Regulators of Appetite

Obesity drugs to regulate appetite are initially approved by the FDA but were removed (fenfluramine, dexfenfluramine, phenylpropylamine) or voluntarily withdrawn (sibutramine) because of serious adverse events.

Combination Therapies by the Current Drugs

1) Glycemic Control and Adverse Events

DPP-4 Inhibitor (Sitagliptin) *vs.* TZD (Pioglitazone) add-on to Metformin

The recent randomized, parallel group study of 16 weeks was designed to compare sitagliptin *vs.* pioglitazone as add-on therapy in patients of T2DM inadequately controlled on metformin alone. Sitagliptin was well tolerated without any incidence of hypoglycemia. Sitgaliptin 100 mg as an add-on to metformin is as effective and well tolerable as pioglitazone 30 mg. There was a significant reduction in the mean body weight and BMI in patients with sitagliptin 100 mg with metformin more than 1,500 mg in contrast to the significant increase in the same in pioglitazone 30 mg with metformin more than1,500 mg. Sitagliptin as an add-on to metformin is as effective and well tolerated as pioglitazone, but does not cause edema or weight gain. In view of its efficacy and good tolerability profile, sitagliptin may be a useful addition to the existing therapeutic armamentarium for treatment of patient with T2DM. The larger and longer studies are needed for validating the results regarding the non-inferiority of DPP-4 inhibitors *versus* TZDs as add-on therapy to metformin [90].

DPP-4 Inhibitor (Sitgaliptin) *vs.* GLP-1 Receptor Agonist (Liraglutide) add-on to Metformin

In the parallel-group, open-label trial, participants (aged 18 -80 years) with T2DM who had inadequate glycemic control on metformin were enrolled. Greater lowering of mean HbA1c was achieved with 1. 8 mg liraglutide (-1. 5 %, 95 % CI -1.63 to -1.37, n=218) and 1. 2 mg liraglutide (-1.24 %, -1.37 to -1.11, n=221) than with sitagliptin (-0. 9 %, -1.03 to -0.77, n=219). Frequency of nausea was more in patients given liraglutide (27 % on 1.8 mg, 21 % on 1. 2 mg) than those given sitagliptin (5 %). There was minor hypoglycemia in about 5 % of subjects in each group. Therefore, liraglutide was superior to sitagliptine regarding the HbA1c decline with low hypoglycemic risk [91].

DPP-4 Inhibitor (Vildagliptin) *vs.* Placebo with and without SU (Glibenclamide)

In a double-blind, four way crossover study, vildagliptin (100 mg) or placebo, with and without glibenclamide (5 mg), was given in random order in 16 healthy men 30 minutes before 75 g oral glucose tolerance test. Hypoglycemia caused by glibenclamide is not enhanced by the vildagliptin co-administration. This may be attributable to a negative feedback mechanism of GLP-1 and GIP secretion limiting the degree of enhanced incretin levels [92].

GLP-1 Receptor Agonist (Liraglutide *vs.* Exenatide) add-on to Oral Hypoglycemic Therapy (LEAD-6)

In a 26-week open-label, parallel-group, multinational study, inadequately controlled T2DM adults were classified by previous oral anti-diabetic therapy and randomly assigned to receive additional liraglutide 1. 8 mg once a day (n=233) or exenatide 10 μg twice a day (n=231). The primary outcome was HbA1c change. Intention to treat was used for efficacy analyses. Liraglutide provided significantly greater improvements in glycemic control than did exenatide (estimated treatment difference of mean HbA1c; -0.33, 95 % CI -0.47 to -0.18, $P < 0.0001$, patients achieved a HbA1c value of less than 7 %; OR 2.02, 95 % CI 1.31 to 3.11, $P = 0.0015$, estimated treatment difference of mean fasting plasma glucose; -1.01 mmol/L, 95 % CI -1.37 to -0.65, $P < 0.0001$). Both drugs were well tolerated, but nausea was less persistent (estimated treatment rate ratio; 0.448, $P < 0.0001$) and minor hypoglycemia less frequent with liraglutide than with exenatide (rate ratio 0.55, 95 % CI 0.34 to 0.88, $P = 0.0131$). Therefore, liraglutide was generally better tolerated [93].

Long Acting Insulin (Glargine) *vs.* DPP-4 Inhibitor (Sitagliptin) add-on to Metformin (EASIE Study)

Insulin glargine *vs.* sitagliptin in patients whose disease was uncontrolled with metformin. 515 patients were randomly assigned to glargine (n=250) or sitagliptin (n=265). Reduction of HbA1c was greater for patients on glargine (-1.72 %

standard error [SE] 0.06) than for those on sitagliptin (-1.13 %, SE 0.06). The estimated rate of symptomatic hypoglycemic episodes was greater with glargine than with sitagliptin (4.21, SE 0.54 *vs.* 0.50, SE 0.09). Severe hypoglycemia occurred in 3 patients on glargine and one on sitagliptin [94].

Long Acting Insulin (Glargine) *vs.* Placebo on GLP-1 Receptor Agonist (Liraglutide) add-on to Metformin and SU

The randomized, parallel-group, controlled 26 week trial of 581 patients with T2DM on prior monotherapy and combination therapy was conducted in 107 centers in 17 countries. The primary endpoint was HbA1c. Liraglutide added to metformin and glimepirid provided significantly greater improvement in glycemic control (reduced HbA1c) compared with glargine (1.33 % *vs.* 1.09 %; -0.24% difference, 95 % CI 0.08 to 0.39; p = 0.0015) or placebo (-1.09 % difference, 95 % CI 0.90 to 1.28; p < 0.0001). There was greater weight loss with liraglutide *vs.* placebo (-1. 39 kg, 95 % CI 2.10 to 0.69; p = 0.0001), and *vs.* glargine (-3. 43 kg, 95 % CI 4.00 to 2.86; p < 0.0001) [95].

2) Effect on Morbidity and Mortality

Whether intensive glycemic control decrease macro-vascular events and all-cause mortality in T2DM subjects was unclear in contrast to the substantial benefits to micro-vascular complications. Individual trials have failed to show consistent beneficial effects on cardiovascular outcomes. The following studies including relevant meta-analyses or trial sequential analysis of adequate study selection were undertaken to determine whether intensive treatment and specified therapies are beneficial.

Meta-Analysis of 5 Trials [96]

Predefined inclusion criteria in this meta-analysis required clinical trials to: (1) randomly assign individual with T2DM either to an intensive lowering of glucose *vs.* a standard regimen; (2) measure outcome with a primary endpoint based on cardiovascular events; and (3) be done in stable individuals only, which excluded studies in an acute hospital setting. Five prospective randomized controlled trials of 33,040 participants finally fulfilled the above selection criteria and are included in the meta-analysis. Selected studies are UKPDS (United Kingdom Prospective

Diabetes Study) [97], PROactive (PROspective pioglitAzone Clinical Trial In macroVascular Events) [98-100], ADVANCE (Action in Diabetes and Vascular Disease-PreterAx and DiamicroN MR Controlled Evaluation) [101], ACCORD (Action to Control Cardiovascular Risk in Diabetes Study) [102] and VADT (Veterans Affairs Diabetes Trial) [103]. Intensive treatment included SUs, metformin, TZD, glinide, αGI, insulin, or their combination. A standard regimen was placebo, standard care, or glycemic control of reduced intensity. In this study, OR and 95% CIs were calculated from raw data of every trials because 3 studies (UKPDS, PROactive, and ADVANCE) showed hazard ratios and 95 % CIs but 2 studies (VADT and ACCORD) showed absolute numbers of events. Ray *et al.* studied the effect of intensive glycemic control *vs.* standard therapy on the outcomes of interest with a random-effects-model meta-analysis. Statistical heterogeneity across trials was assessed with χ^2 (P<0.1) and I^2 statistics. At follow-up, intensive therapy group had a mean HbA1c level of 0.9 % (95 % CI 0.88 to 0.92) lower than usual therapy group. During about 163,000 person-years of follow-up, authors recorded 1,497 events of non-fatal myocardial infarction, 2318 of coronary heart disease, 1,127 of stroke, and 2892 of deaths from any cause. Intensive glucose-lowering therapy significantly decreased non-fatal events of myocardial infarction by 17 % (OR 0.83, 95 % CI 0.75 to 0.93) and events of coronary heart disease by 15 % (OR 0.85, 0.77 to 0.93) without heterogeneity (I^2 0.0 %). However, intensive therapy did not significantly affect stroke or all-cause mortality. Heterogeneity was not significant for stroke (I^2 0.0 %), but was high for all-cause mortality (I^2 58 %, P= 0.049). Therefore, an effect on all-cause mortality could not be further clarified without access to individual participant data. In contrast, strong evidence suggests that therapy for dyslipidemia and high blood pressure benefit all-cause mortality in T2DM subjects (9 % and 27 % reduction, respectively) [104-107].

Intensive therapy group suffered a more frequent hypoglycemia than standard therapy group (weighted averages 38.1 % *vs.* 28.6 %). Overall, severe hypoglycemia was much less common than was hypoglycemia, but subjects given intensive therapy had a severe hypoglycemia almost twice as many compared with those on standard regimens (weighted averages 2.3 % *vs.* 1.2 %). At the end of the study, subjects given intensive therapy were 2.5 +/- 1.2kg (mean +/-SD) heavier than those on standard regimens. In the participants of two studies (ACCORD and VADT) with increased mortality in the intensive treatment group, duration since T2DM diagnosis at baseline (>10 years) was longest; HbA1c level at baseline was highest; and a risk of hypoglycemia was greater. In the ACCORD study, risk of cardiovascular death and non-coronary cardiovascular death was significantly higher. In ACCORD, HbA1c was reduced by 1.5 % within 6 months and average HbA1c level was less than 6 % by 1 year in intensive therapy group through early and aggressive use of insulin. In addition, a greater proportion of intensive therapy group took rosiglitazone at the end of follow-up (92 %) compared with those taking standard regimens (58 %). In contrast, ADVANCE study showed HbA1c concentration reduction by only 0.5 % within 6 months and much slower (about 36 months) achievement of target HbA1c level of 6.5 % or less with much lower use of longer acting insulin. Taken together, a practical clinical approach might reduce HbA1c level steadily to less stringent targets with care taken to avoid severe hypoglycemia.

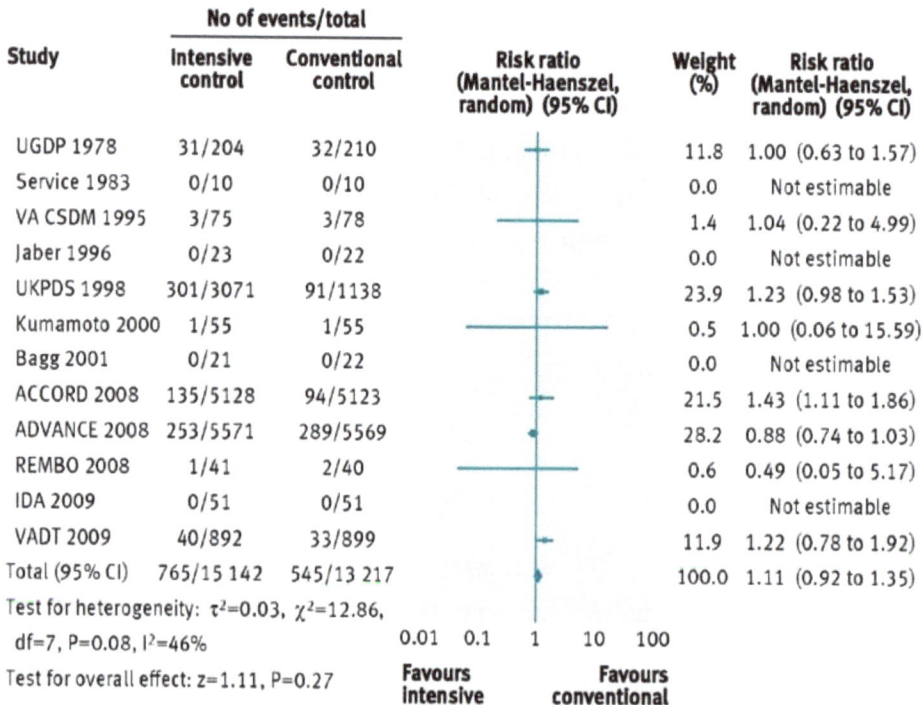

Study	No of events/total		Risk ratio (Mantel-Haenszel, random) (95% CI)	Weight (%)	Risk ratio (Mantel-Haenszel, random) (95% CI)
	Intensive control	Conventional control			
UGDP 1978	31/204	32/210		11.8	1.00 (0.63 to 1.57)
Service 1983	0/10	0/10		0.0	Not estimable
VA CSDM 1995	3/75	3/78		1.4	1.04 (0.22 to 4.99)
Jaber 1996	0/23	0/22		0.0	Not estimable
UKPDS 1998	301/3071	91/1138		23.9	1.23 (0.98 to 1.53)
Kumamoto 2000	1/55	1/55		0.5	1.00 (0.06 to 15.59)
Bagg 2001	0/21	0/22		0.0	Not estimable
ACCORD 2008	135/5128	94/5123		21.5	1.43 (1.11 to 1.86)
ADVANCE 2008	253/5571	289/5569		28.2	0.88 (0.74 to 1.03)
REMBO 2008	1/41	2/40		0.6	0.49 (0.05 to 5.17)
IDA 2009	0/51	0/51		0.0	Not estimable
VADT 2009	40/892	33/899		11.9	1.22 (0.78 to 1.92)
Total (95% CI)	765/15 142	545/13 217		100.0	1.11 (0.92 to 1.35)

Test for heterogeneity: τ^2=0.03, χ^2=12.86, df=7, P=0.08, I^2=46%

Test for overall effect: z=1.11, P=0.27

0.01   0.1   1   10   100

Favours intensive   Favours conventional

Figure 2: Forest plot for cardiovascular mortality compared between intensive and conventional glycemic control (Reproduced from Ref. [108]).

Meta-Analysis and Trial Sequential Analysis (TSA) of 14 Trials [108]

Systematic review with meta- analysis and TSA [109, 110] of randomized trials were designed to evaluate the effect of targeting intensive glycemic control compared with conventional glycemic control on all-cause as well as cardiovascular mortality, non-fatal myocardial infarction, micro-vascular complications, and severe hypoglycemia in T2DM patients. Potentially relevant 93 articles were identified from 10,047 articles searched by databases, and then 51 references including 14 trials were finally selected by several exclusion criteria. From risk of bias assessments among 14 trials, most adequate trials were UGDP (University Group Diabetes Program), UKPDS, ADVANCE, ACCORD, VADT and IDA (Insulin Diabetes Angioplasty) [111, 112] among which 4 trials were the same as the above report [96]. This study used both a random effects model and a fixed effect model. Heterogeneity was examined with I^2 statistic. Strength of this study is the meta-analysis included trials published in languages other than English or tested for the risk of having false positive P values or unrealistically narrow CIs and done as a Cochrane systematic review [113]. Meta-analysis showed no significant effect of intensive glycemic control on all-cause mortality (relative risk [RR]; 1.02, 95% CI 0.91 to 1.13, P=0.74, 28,359 participants, 12 trials) compared with conventional glycemic control. The meta-analysis of the same 12 trials did not show a statistically significant effect of the intensive glycemic control on cardiovascular mortality (RR; 1.11, 0.92-1.35) (Fig. **2**), even though intensive glycemic control showed significant effect on cardiovascular mortality in UGDP study (RR; 2.66, 1.13-6.29) and in ACCORD study (RR; 1.43, 1.02-2.02). A total of 1237 non-fatal myocardial infarction were recorded in 28,111 participants. The effect estimate showed a significant benefit of targeting intensive glycemic control in a conventional meta-analysis (RR; 0.85, 0.76 to 0.95, P=0.004). However, TSA did not show sufficient evidence of a benefit (TSA adjusted 95 % CI 0.71 to 1.02). In micro-vascular complications, the effect estimate on retinopathy showed a significant benefit in favor of intensive glycemic control (RR 0.80; 0.67 to 0.94, P=0.009), while TSA showed a lack of sufficient evidence (TSA adjusted 95 % CI 0.54 to 1.17). No statistically significant effect of intensive glycemic control on nephropathy was found (RR; 0.83; 0.64 to 1.06, P=0.13). Taken together, the meta-analysis with TSA provided insufficient evidence to confirm or exclude that intensive glycemic control favors for reductions not only of micro-vascular complications but also of macro-vascular events and mortality in patients with T2DM. However, meta-analysis of intensive *vs.* conventional control clearly confirmed a significant effect on severe hypoglycemia (RR; 2.39, 1.71-3.34, P<0.001), although the definition of severe hypoglycemia varied among trials.

DPP-4 Inhibitors (Linagliptin) *vs.* SU (Glimepiride)

777 patients were randomly assigned to DPP-4 inhibitors, linagliptin and 775 to SU, glimepiride. Fewer participants had hypoglycemia (58 *vs.* 280, P<0.0001), or severe hypoglycemia (1 *vs.* 12) with linagliptin compared with glimepiride, although reductions in adjusted mean HbA1c were similar. Linagliptin was associated with fewer cardiovascular events (12 *vs.* 26; RR 0.46, 95 % CI 0.23-0.91, P=0.0213) and fewer non-fatal stroke (3 *vs.* 11; RR 0.27, 95 % CI 0.08-0.97, P=0.03) [114].

Nateglinide and Valsartan in Impaired Glucose Tolerance Outcomes Research (NAVIGATOR Study)

Among 9,306 persons with impaired glucose tolerance (fasting plasma glucose higher than 95 and less than 126 mg/dl) and established cardiovascular disease or cardiovascular risk factors, 4,645 were randomly assigned to receive nateglinide 60 mg (initial dose; 30 mg) three times daily and 4,661 to receive placebo for 5 years. Nateglinide did not decrease the prevalence of diabetes or the co-primary composite cardiovascular events as compared with placebo [115].

Basal Supported Oral Therapy (BOT), Basal Plus or Basal Bolus [Outcome Reduction with an Initial Glargine Intervention (ORIGIN trial)]

The aim of this study is to clarify whether the provision of sufficient basal insulin to normalize fasting plasma glucose levels may decrease cardiovascular outcomes. Subjects (mean age, 63.5 years, n=12,537) with cardiovascular risk factors and impaired fasting glucose as well as glucose tolerance, or T2DM, were randomly assigned to receive insulin glargine to achieve fasting plasma glucose level below 95 mg per deciliter or standard care and to receive n-3 fatty acids or placebo with the use of 2- by -2 factorial design. When used to achieve normal fasting plasma glucose levels for more than 6 years, cardiovascular event rates were similar in the glargine and standard-care groups (2.94 and 2.85 per 100 person-years, respectively). Glargine reduced new-onset DM, but it also caused an increase in hypoglycemia and a modest weight gain by 1.6 kg [116].

Insulin *vs.* Other Anti-Hyperglycemic Therapies

A retrospective cohort study was performed using data from the UK General Practice Research Data Base, 2000-2010. Primary care patients with T2DM (n=84,622) were treated with one of five glucose-lowering regimens: metformin, SU mono-therapy, insulin mono-therapy, metformin + SU, and insulin + metformin combination therapy. Main outcome was the risk of the first major cardiac event, first cancer, or mortality. The adjusted hazard ratio for the primary end point and all-cause mortality were higher for insulin mono-therapy *vs.* all other regimens (1.808, 95 % CI 1.630 to 2.005, P<0.0001, 2.197, 1.983 to 2.434, P<0.0001, respectively). The adjusted hazard ratio for one of the secondary end points, neuropathy was higher for insulin mono-therapy *vs.* all other regimens (2.146, 95 % CI 1.832 to 2.514, P<0.0001) [117].

2. THE EMERGING MEDICATIONS FOR T2DM AND OBESITY (Fig. 3)

Despite the number of medications for T2DM, many subjects with T2DM do not achieve good glycemic control. Some present drugs for T2DM have adverse events such as weight gain or hypoglycemia. T2DM is a progressive disease, and most patients need combination therapy to control hyperglycemia.

1) Salt Glucose Cotransporter 2 (SGLT2) Inhibitors

Glucose is normally filtered in the glomerulus of the kidney and is reabsorbed in the proximal tubules. Glycosuria occurs when blood glucose levels exceed the renal threshold of glucose (approximately 10 mmol/l (160-180 mg /dl). In total, 98 % of the urinary glucose is transported across the membrane of the proximal tubule by SGLT2, a high-capacity, low-affinity glucose transporter. A natural mutation in the SLC5A2 gene result in a SGLT2 defect with significant glycosuria. Individuals with this mutation have not complained glycosuria-related problems, such as urinary tract infections [118]. Therefore a therapeutic option in T2DM is to mimic the effect of the SLC5A2 mutation and prevent the reabsorption of renal-filtered glucose, thereby decreasing hyperglycemia, without weight gain or hypoglycemia [119]. Systematic review was performed to assess the clinical effectiveness and safety of the SGLT2 inhibitors in dual or triple therapy in T2DM [120]. Seven trials, published in full, assessed dapagliflozin and one assessed canagliflozin. Trial quality level appeared high.

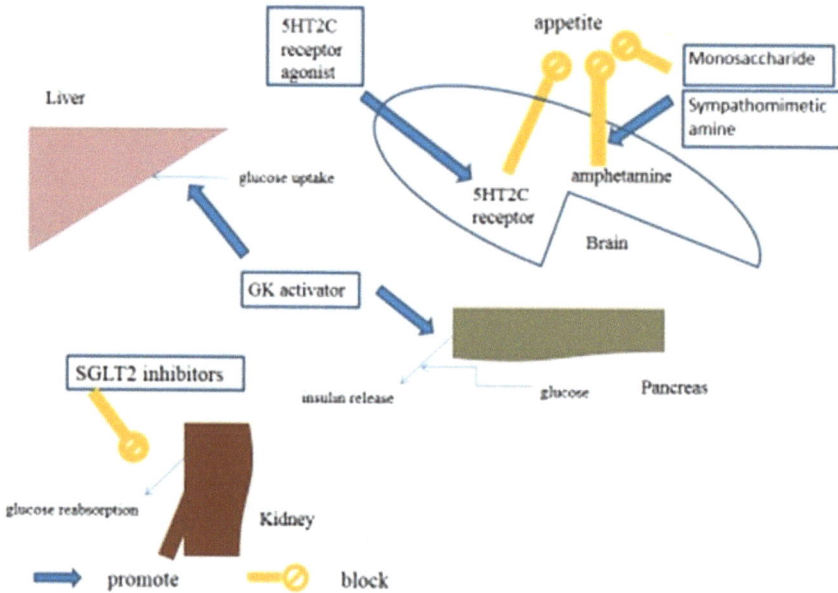

Figure 3: Major target organs and actions of new emerging drugs in T2DM and obesity SGLT2: salt glucose cotransporter 2, GK: glucokinase, 5HT2c: 5-hydroxytryptamine 2c.

Dapagliflozin

Dapagliflozin 10 mg decreased HbA1c by -0.54 % (95 % CI; -0.67 to -0.40) compared to placebo, but there was no difference compared to glipizide. Dapagliflozin resulted in weight loss by - 1.81 kg (95 % CI; -2.04 to -1.57 compared to placebo). Long-term trial extensions suggested that effects were maintained over time. More data on cardiovascular, cerebrovascular and cancer safety are needed, with the FDA having concerns about breast and bladder cancers [121]. US FDA approved dapagliflozin in November 2012.

Canagliflozin

Canagliflozin decreased HbA1c slightly more than sitagliptin (up to -0.21 % *vs.* sitagliptin). Canagliflozin resulted in weight loss up to -2.3 kg compared to placebo. Long-term trial extensions suggested that effects were maintained over time. US FDA approved canaglifrozin in March 2013.

Other SGLT2 Inhibitors

Other SGLT2 inhibitors are ipragliflozin, luseogliflozin, empagliflozin and tofogliflozin. US FDA approved empagliflozin in August 2014. In the recent clinical trials of tofogliflozin, the observed mild and moderate adverse events were hyperketonemia, ketonuria, and pollakisuria. The incidence of hypoglycemia was low [122].

2) Glucokinase (GK) Activators

GK is expressed in the liver and pancreas and acts as a glucose-sensing enzyme. This enzyme activation enhances glucose uptake in the liver and insulin release from the pancreas [123]. It should act in only glucose dependent manner and decrease the risk for hypoglycemia. Taken together, it is an ideal target for T2DM treatment [124]. Many GK activators are being developed with promising preclinical results, and some of them are clinically tried [125, 126].

3) Anti-obesity Drugs

Lorcaserin

Lorcaserin (Belviq) is a selective agonist of the serotonin (5-hydroxytryptamine 2c)(5- HT2c) receptor and has been approved in the USA for the obesity treatment in parallel to lifestyle correction in adults of body mass index (BMI) more than 30 kg/m^2, or overweight adults of BMI more than 27 kg/m^2 with at least one weight-related co-morbidity (*e.g.* T2DM, dyslipidemia, hypertension). The efficacy and safety of lorcaserin was examined in a randomized, multi-center, placebo-controlled, double-blind clinical trials (BLOOM-DM [127], BLOSSOM [128] and BLOOM [129]) using 10 mg lorcaserin twice a day. In a pooled analysis, placebo-subtracted change of body weight from baseline was significantly greater in lorcaserin (n=3,098) than in placebo (n=3,038) (-3. 3 kg [95 % CI -3.6, -2.9], p<0.001). Lorcaserin was well tolerated with headache and hypoglycemia, the most frequently reported adverse events in T2DM patients given SU or insulin therapy. Weight regain was not found at least up to the second year of clinical study of lorcaserin [130]. Although it was not established whether lorcaserin was not inferior to placebo, it is unlikely that lorcaserin is associated with the degree of valvular risk defined by FDA that was previously reported with the appetite losers dexfenfluramine and fenfluramine.

Phentermine-Topiramate

Phentermine is a sympathomimetic amine and suppresses appetite through the central nervous system with the mechanism similar to amphetamine [130-132]. In a pooled analysis of several randomized, placebo-controlled studies up to 24 weeks, phentermine caused a 3. 6 kg weight loss compared to placebo [133]. Major adverse events of phentermine taken too late in the day are hypertension, tachyarrhythmias, and insomnia. Topiramate is a monosaccharide with sulfamate-substitution initially approved by the FDA in 1996 to control seizure. Weight loss was noted as adverse events when topiramate was used for seizures, migraine prevention, and bipolar disorder. Bray *et al.* conducted and reported a randomized, double-blind, placebo-controlled, dose-ranging trial in 2003 [134]. Two handreds and forty eight patients completed the trial of 64, 96, 192, or 384 mg/day topiramate through the planned 6 months. BMIs (kg /m^2) were 30-50 in participants, and 27-30 in those with well controlled hypertension or dyslipidemia. In this trial, all the topiramate groups resulted in significantly greater weight loss than the placebo group, although weight loss in the 384 mg/day group did not differ from the 192 mg/day group [134]. By the clinical development, FDA approved a combination of phentermine and topiramate named by Qsymia for the clinical management of obesity in July 2012 [135-140]. This combined preparation of extended release topiramate (topiramate ER) and phentermine can be prescribed once daily. Five major clinical trials are EQUIP, EQUATE, CONQUER, SEQUEL and FORTRESS [139-144]. SEQUEL trial remarkably showed a 76 % reduction of new onset T2DM in the maximum dose treated group. These combination therapy for obesity resulted in successful enhancement of synergism, improvement of efficacy including duration of action and yet simultaneous reduction of overall adverse events.

3. THE FUTURE MEDICATIONS FOR T2DM AND OBESITY (Table 1)

The various research for T2DM and obesity can be categorized into technological and pharmacological approach. The most advanced technology is insulin and glucagon delivery *via* an artificial pancreas, and the system components are already usable, and may be available within 10 years. Technology of pancreas and islet cell transplants are not yet widely available because of a lack of appropriate donor tissue and graft survival after transplant. However, significant progress has been made by additional research [145]. Other cell types such as induced pluripotent stem cells (iPS) can replace β cells as a new option in the near future. These technological approach is definitively required for T1DM rather than T2DM. Pharmacological approach seems the most promising to reduce the T2DM burden. Pharmacological research regarding the agents to selectively restore energy balance is now the most prospective approach to treatments for T2DM and obese subjects. This section focuses on the current pharmacological research for T2DM and obesity.

1) Cyclin-Dependent Kinase 5 (Cdk5) Inhibitor

In the obesity, pro-inflammatory signals induce the p35, Cdk5 partner and targets of numerous cytokines and pro-inflammatory signals, break down to p25, a more active and stable subunit of p35. The p25 is then translocated to the nucleus, where it binds and activates Cdk5 gene. Choi *et al.* [146] show that Cdk5 in turn causes serine 273 phosphorylation of PPARγ. Therefore, Cdk5 inhibitor prevents the specific PPARγ target gene transcription and provides with favorable effects against obesity. PPARγ is the receptor for the TZD, anti-diabetic drugs rosiglitazone and pioglitazone, which they prevent serine 273 phosphorylation. Although these drugs are full classical agonists for this nuclear receptor, many PPARγ-based drugs have a variable biochemical activity and prevent the obesity-related PPARγ phosphorylation by Cdk5 [147]. The same authors reported novel synthetic compounds having a unique mode of binding to PPARγ without agonistic action in classical transcription and blocking the Cdk5-induced phosphorylation in cultured adipocytes and in insulin-resistant mice [147]. Furthermore, one of novel compounds, SR1664, has potent effect against T2DM without serious adverse events like the fluid retention and weight gain frequently observed in PPARγ drugs. SR1664 also does not affect bone formation *in vitro* unlike TZDs. These data indicate that novel classes of drugs against T2DM can be generated by specifically targeting the Cdk5-mediated PPARγ phosphorylation.

Table 1: Candidate compounds for T2DM and obesity treatment under pharmacological research

Compounds	Target cells or organ	Action
Cdk5 inhibitor Leptin	adipocytes skeletal muscle	prevent PPARγ phosphorylation on serine 273 increase glucose uptake *via* hypothalamic sympathetic nervous system fatty acid oxidation *via* AMPK
Adiponectin mimetics	skeletal muscle liver	activate AMPK pathway by AdipoR1 activate PPARγ pathway by AdipoR2
SIRT1 activator	adipose tissues skeletal muscle liver	improve insulin sensitivity, lower plasma glucose increase mitochondrial capacity
NPY2R antagonist	adipose tissues	prevent proliferation and differentiation of fat cells prevent angiogenesis
11β-HSD1 inhibitor IL-6	skeletal muscle pancreas β cells small intestine L cells	prevent glucocorticoid-induced insulin resistance increase GLP-1 production
Betatrophin	pancreas β cells	β cell proliferation and β cell mass expansion improve glucose tolerance
Chemerin	adipocytes pancreas β cells	increase differentiation and lipolysis enhance glucose-stimulated insulin secretion
DLPC	liver (LRH-1 ligand)	increase bile acid biosynthetic enzymes lower hepatic TG, and serum glucose decrease hepatic steatosis improve glucose homeostasis
DNJ	small intestine liver	inhibit glucose absorption by down-regulation of SGLT1, Na+/K+-ATPase, and GLUT2 increase hepatic glycolysis enzymes decrease gluconeogenesis enzymes
FGF21	adipocytes liver	enhance lipolysis, β oxidation, glucose uptake decrease glucose production
FGF1	liver, muscle	increase insulin sensitivity
SCD1 inhibitor	liver	inhibit biosynthesis of triglyceride and cholesterol ester and conversion of saturated fatty acid to monounsaturated fatty acid

Cdk5: cyclin-dependent kinase 5, 11β-HSD1: 11β-hydroxysteroid dehydrogenase type 1
NPY2R: neuropeptide Y2 receptor, DLPC: dilauroyl phosphatidylcholine, DNJ: 1-deoxynojirimycin FGF: fibroblast growth factor, SCD1: stearoyl Co enzyme A desaturase-1

2) Leptin

Leptin is an adipocyte derived hormone playing a pivotal role in regulation of food intake, energy expenditure, and neuroendocrine function. Leptin is known to regulate glucose and lipid metabolism in peripheral tissues independent of the anorexic effect in rodents and humans. Leptin is shown to ameliorate diabetes in rodents and subjects with lipodystrophy. Leptin increases glucose uptake through the axis of hypothalamic-sympathetic nervous system and ß-adrenergic system, and fatty acid oxidation *via* AMPK in skeletal muscle of rodents *in vivo* [148, 149]. Leptin-induced fatty acid oxidation decreases muscle lipid accumulation, causing "lipotoxicity". Leptin activates AMPK directly on muscle and indirectly on the medial hypothalamus-sympathetic nervous system and α-adrenergic mechanism. Therefore, leptin plays a pivotal role in the regulation of glucose and fatty acid metabolism in skeletal muscle. Leptin therapy is not successful for T2DM subjects with leptin resistance but beneficial for those with low adipose mass and higher leptin-sensitivity [150]. Leptin may have a therapeutic usefulness for lipodystrophic and non-lipodystrophic insulin-resistant and T2DM subjects [151].

3) Adiponectin Mimetics

Adiponectin is a hormone derived from adipocytes and plays an important role in the prevention of metabolic derangements causing T2DM, obesity, and atherosclerosis. Up-regulation of adiponectin or its receptor has been shown to be a number of therapeutic benefits. Adiponectin binds to receptors including adiponectin receptor 1 (AdipoR1) and adiponectin receptor 2 (AdipoR2). AdipoR1 mainly affects the AMPK pathway and AdipoR2 mainly affects the PPARγ pathway and this correlates to plasma insulin levels *in vivo* [152, 153]. Since it is difficult to convert the full size adiponectin into a viable drug, adiponectin receptor agonists were identified by high-throughput screening. The fluorescence polarization assay was conducted to screen against larger small molecular compound libraries. A natural product library containing 10,000 compounds was screened by Sun *et al.* and 9 compounds were selected for validation [154]. These compounds were confirmed their agonistic activity by the second-step *in vitro* tests. The most active AdipoR1 agonists are matairesinol, arctiin, (-)-arctigenin and gramine. The most active AdipoR2 agonists are parthenolide, taxifoliol, deoxyschizandrin, and syringin. These compounds may be beneficial drug candidates for hypoadiponectin related diseases such as obesity and T2DM.

4) SIRT1 Activators

An NAD⁺-dependent deacetylase SIRT1 mainly modulates downstream pathways of calorie restriction with favorable effects on glucose metabolism and insulin sensitivity [155-161]. Milne *et al.* reported small SIRT1 activator molecule, resveratrol that was not structurally related to SIRT1 but 1,000-fold more potent [162]. Resveratrol binding to the SIRT1—peptide substrate complex at an allosteric site amino-terminal to the catalytic domain, caused the higher affinity for acetylated substrates. In diet-induced and genetically obese mice, these compounds improve insulin sensitivity with decreased plasma glucose and increased mitochondrial function. Hyperinsulinemic-euglycemic clamp studies indicate that SIRT1 activators ameliorate whole-body glucose metabolism and insulin sensitivity in skeletal muscle, adipose tissue, and liver of Zucker fa/fa rats. Therefore, SIRT1 activator is a prospective new approach for T2DM treatment.

5) Selective Antagonists for Neuropeptide Y (NPY) 2 Receptor (NPY2R)

NPY and NPY2R are involved in the high calorie diet plus stress-induced visceral obesity in mice by enhanced proliferation and differentiation of fat cells and angiogenesis in the adipose stromal tissues [163]. Stress induces catechol-amines and glucocorticoid secretion, thereby stimulating NPY secretion from sympathetic nerve terminals and then acting *via* NPY2R in adipose tissues [163]. NPY not only stimulates food intake by stress through the hypothalamic NPY1R [164-166] but also directly accumulates fat in visceral adipose tissue *via* NPY2R. A leucine (7) to proline (7) polymorphism in the signal peptide of NPY has been reported to associate with birth weight, increased serum low-density lipoprotein cholesterol levels, serum TG levels, atherosclerosis, and T2DM [167-169]. On the other hand, obesity has been reported to significantly associate with 5 single nucleotide polymorphisms (SNPs) of NPY2R gene in Europe, although minor allele frequency of these SNPs is high in Asia [170-174]. Among them, 2 SNPs, rs6857715 and rs6857530, are located on the 5'-flanking region of the NPY2R gene, suggesting the possible regulation of NPY2R gene transcription. We examined the association between these SNPs and metabolic parameters. We have reported significant associations between these SNPs and serum HDL cholesterol levels in Japanese men [175].

6) 11β-hydroxysteroid Dehydrogenase type 1 (11β-HSD1) Inhibitor

11β-HSD1 is an enzyme that converts from inactive glucocorticoid cortisone to active cortisol. IRS1 phosphorylation at Ser 307 residue is involved in glucocorticoid-induced insulin resistance. Therefore, the reduction of glucocorticoid availability by selective 11β-HSD1 inhibitors is important to improve insulin resistance predominantly caused by impaired lipid metabolism in skeletal muscle [176]. Clinical trials for obese and T2DM patients using 11β-HSD1 inhibitors are emerging [177]. Their efficacy in muscle may be benefitial in the therapy of T2DM and insulin resistance. Furthermore, net hepatic glucose output and endogenous glucose production were significantly decreased by 11β-HSD1 inhibition due to a significant decrease in net hepatic glycogenolysis without effect on gluconeogenic flux compared with placebo in dogs [178].

7) Interleukin-6 (IL-6)

Plasma concentrations of IL- 6 are elevated in exercise, obesity and T2DM [179-181]. Exogenous IL-6 injection or endogenous IL-6 in response to exercise stimulate GLP-1 release from intestinal L cells and pancreatic β cells, thereby improving insulin release and hyperglycemia. The mechanism by which IL-6 increased GLP-1 secretion from β cells is through increased pro-glucagon (which is encoded by GCG) and prohormone convertase 1/3 expression [182]. In T2DM models, the beneficial effects of IL-6 were maintained, and neutralization of IL-6 caused further hyperglycemia and GLP-1 reduction from pancreas. Therefore, IL-6 is involved in crosstalk between insulin acting tissues and GLP-1 producing intestinal L cells and pancreatic islets to adapt to insulin needs, suggesting that drugs modulating this new loop may be useful in T2DM.

8) Betatrophin

Pancreatic β cell mass generation is beneficial for treating T2DM. Pancreatic β cells are generated primarily by self-duplication in adults. Yi *et al.* have reported remarkable pancreatic β cell proliferation and β cell mass expansion on a mouse insulin resistance model [183]. They identified betatrophin predominantly expressed in liver and fat of this model. Betatrophin is a protein of 198 amino acids and its gene consists of 4 exons and locates in the intron of another gene,

Dock6, on the opposite strand. Betatrophin is highly conserved in all mammalians but not in non-mammalians. Expression of betatrophin correlates with proliferation of β cell in other insulin resistant mouse models and during gestation. Transient expression of betatrophin in mouse liver significantly and specifically stimulates β cell proliferation as well as β cell mass expansion, and ameliorates glucose tolerance. Therefore, betatrophin administration could replace insulin treatments by increasing the β cell mass in T2DM subjects. Such treatments could provide the long-term normal glycemic control as a remarkable curative therapy in T2DM subjects.

9) Chemerin

Chemerin was originally identified in the skin as tazarotene-induced gene 2 [184] and subsequently as the ligand for a G protein-coupled receptor, ChemR23 /CMKLR1. This receptor is similar to chemokine receptor and is expressed in immature dendritic cells and macrophages [185]. Chemerin promotes chemotaxis of these cells [185], suggesting the association of chemerin with a proinflammatory state. Several investigators have reported that chemerin is an adipokine [187-190]-regulating adipocyte differentiation and lipolysis [186, 187]. Moreover, chemerin stimulates insulin-dependent glucose uptake in adipocytes [188]. These data suggest that chemerin plays a regulatory role as an adipokine in metabolism. Indeed, serum chemerin levels correlate with BMI, fasting serum glucose, insulin, TG, and cholesterol levels [187]. However, the physiological role of chemerin remained to be clarified. We have demonstrated the glucose intolerance in chemerin-deficient mice even though macrophage accumulation was reduced in adipose tissue [190]. This glucose intolerance was mainly caused by increased glucose production in the liver and impaired insulin secretion from the pancreas. As chemerin and its receptor ChemR23 were expressed in β-cell, we have performed experiments using isolated pancreatic islets and perfused pancreas and demonstrated impaired glucose-stimulated insulin secretion (GSIS) in chemerin-knockout mice. In contrast, enhanced GSIS and ameliorated glucose tolerance were observed in chemerin transgenic mice. These results suggest that chemerin regulates β-cell function and plays a pivotal role in glucose metabolism in a tissue-dependent manner, although the latter role is still controversial.

10) Dilauroyl Phosphatidylcholine (DLPC)

Nuclear hormone receptors are involved in a variety of metabolic pathways. The orphan nuclear receptor LRH-1 (NR5A2) mediates bile acid synthesis [191, 192]. Phospholipids as potential LRH-1 ligands were structurally identified [193-195], but their functional significance was unclarified. Lee *et al.* have reported DLPC, an unusual phosphatidylcholine species with 2 saturated 12 carbon acyl side chains acting as an agonistic ligand for LRH-1 *in vitro* [196]. DLPC induces liver enzymes for bile acid synthesis with increases in bile acid levels, and reduces hepatic TG and serum glucose. DLPC also reduces liver steatosis and ameliorates glucose metabolism in two mouse models of insulin resistance. Both the anti-diabetic and lipotropic effects are not found in liver-specific Lrh-1 deficient mice. These findings suggest that an LRH-1 dependent phosphatidylcholine signaling pathway could regulate bile acid and glucose metabolism.

11) 1-deoxynojirimycin (DNJ)

There are several evidences indicating that changes of the brush border membrane (BBM) and basolateral membranes (BLM) in the small intestine caused the increased glucose absorption in T2DM [197]. These changes are mainly due to increased activity, mRNA and protein levels of sodium glucose transport protein (SGLT1), Na+/K+-ATPase and glucose transporter 2 (GLUT2) [198-200]. Longstanding deterioration of endocrine regulation by supra-physiological levels of glucose absorption further induces the metabolic derangements by changes in the activities of key enzymes GK, phosphofructokinase (PFK), pyruvate kinase (PK), phosphoenolpyruvate carboxykinase (PEPCK) and glucose-6-phosphatase (G-6-Pase). These changes decrease peripheral glucose utilization and enhance hepatic glucose production [201, 202]. Therefore, suppression of glucose absorption at BBM or BLM and/or modulation of the hepatic enzyme activities for carbohydrate metabolism provide a beneficial effect on the improvement of hyperglycemia in T2DM subjects. DNJ is derived from Mulberry leaves and potently suppresses intestinal α-glycosidases. In the report by Li *et al.* studying the role of DNJ on glucose absorption and metabolism in normal and diabetic mice [203], oral and intravenous glucose tolerance tests and labeled $^{13}C_6$-glucose uptake assays resulted in DNJ suppression of glucose absorption from the small intestine. They also demonstrated the decrease by DNJ of SGLT1, Na+/K+-

ATPase and GLUT2 mRNA and protein levels in the intestine. Pretreatment with 50 mg/kg DNJ increased the protein levels of hepatic glycolytic enzymes (GK, PFK, PK, PDE1) and decreased the expression of gluconeogenesis enzymes (PEPCK, G-6-Pase). These results suggest that DNJ suppresses glucose absorption from the small intestine and enhances glucose metabolism in the liver by directly regulating the protein levels involved in glucose transport systems, and enzymes for glycolysis as well as gluconeogenesis.

12) Fibroblast Growth Factor (FGF) Family Members, FGF21, FGF1

FGF family member, FGF21 is produced in liver, white and brown adipose tissue, and pancreas. FGF21 has physiological multi-functions such as increased thermogenesis, lipolysis, glucose uptake, ketogenesis, fatty acid oxidation and decreased fatty acid sysnthesis particulary under stress. Administration of FGF21 also has shown to sensitize insulin action through reduced hepatic glucose production, increasing glucose uptake in adipocytes without major adverse events in rodents [204, 205]. FGF21 mimetics showed some reduction in body weight but effects on glucose metabolism was modest in clinical trials [206]. The other FGF family member, FGF1 is also regulated its transcription by PPARγ in adipocytes. Fgf1 knockout mice revealed severe insulin resistance under stress and high fat diet [207, 208]. Recent report by Suh *et al.* has shown that acute and chronic administration of recombinant FGF1 resulted in lowering glucose levels and sensitized insulin action through liver and muscle without side effects in diabetic mice [209]. Therefore, non-mitogenic action of these FGF family may be promising candidate for safe insulin sensitizer as T2DM therapeutics.

13) Stearoyl-Coenzyme A Desaturase-1(SCD1) Inhibitors

SCD1 plays a vital role in conversion of saturated fatty acids (SFAs) to monounsaturated fatty acids. Inhibition of SCD1 has been reported to result in protection against diet-induced obesity, fatty liver and insulin resistance in metabolic syndrome rodent models [210, 211]. These findings suggest that SCD1 inhibitors are promising candidates for T2DM and obesity. On the other hand, proinflammatory SFAs accumulated by SCD1 inhibition promote inflammation including atherosclerosis, so further studies are required to examine whether these compounds are well tolerated in inflammatory context.

CONCLUSION

T2DM and obesity are lifestyle-related diseases but often need pharmacological intervention. Unlike SU and insulin, metformin and incretin-based therapies are useful for glycemic control without severe hypoglycemia and weight gain. Long term prospective studies are underway to determine whether incretin-based therapies are safe and also favorable for reduction of vascular complications and mortality. Meta-analysis and trial sequential analysis of randomized clinical trials indicate that intensive glycemic control is not necessarily favorable to prevent vascular complications and mortality compared with conventional control. In the near future, new drugs including those presented here will be developing and expected to be safe and useful for glycemic control to ameliorate morbidity and mortality in clinical practice of T2DM and obesity.

ACKNOWLEDGEMENTS

This work was supported from Grant-in-Aid for Scientific Research (C), Japan Society for the Promotion of Science (JSPS). Disclosure Summary: The author has nothing to declare.

CONFLICT OF INTEREST

The authors confirm that this chapter contents have no conflict of interest.

ABBREVIATIONS

11β-HSD1 = 11β-hydroxysteroid dehydrogenase type 1

5-HT2c   = 5-hydroxytryptamine 2c

α-GI   = α-glucosidase inhibitors

AdipoR   = adiponectin receptor

AMPK   = adenosine monophosphate-activated protein kinase

BBM   = brush border membrane

BLM = basolateral membranes

BMI = body mass index

BOT = basal supported oral therapy

Cdk5 = cyclin-dependent kinase 5

CI = confidence intervals

DLPC = dilauroyl phosphatidylcholine

DNJ = 1-deoxynojirimycin

DPP-4 = dipeptidyl peptidase 4

e.g. = exempli gratia

FGF = fibroblast growth factor

G-6-Pase = glucose-6-phosphatase

GIP = gastric inhibitory polypeptide

GK = glucokinase

GLP-1 = glucagon like peptide- 1

GLUT2 = glucose transporter 2

GSIS = glucose-stimulated insulin secretion

HDL = high-density lipoprotein

IL-6 = interleukin-6

K_{ATP} = ATP-sensitive potassium

NASH = non-alcoholic steatohepatitis

NPY   = neuropeptide Y

NPY2R   = neuropeptide Y2 receptor

ORs   = odds ratios

PDE1   = phosphodiesterase type 1

PEPCK   = phosphoenolpyruvate carboxykinase

PFK   = phosphofructokinase

PK   = pyruvate kinase

PPAR-γ = peroxisome proliferator-activated receptor-γ

q.i.d.   = quarter in die (four time a day)

RR   = relative risk

SCD1   = stearoyl-coenzyme A desaturase-1

SE   = standard error

SFAs   = saturated fatty acids

SGLT2   = salt glucose cotransporter 2

SIRT   = sirtuin

SNPs   = single nucleotide polymorphisms

SU   = sulfonylurea

SUR1   = SU receptor 1

T2DM   = type 2 diabetes mellitus

TG   = triglyceride

TZDs = thiazolidinediones

US FDA = United States Food and Drug Administration

WMD = weighted mean differences

FULL NAMES OF CLINICIAL TRIALS

ACCORD	=	Action to Control Cardiovascular Risk in Diabetes
ADVANCE	=	Action in Diabetes and Vascular disease: Preterax and Diamicron MR Controlled Evaluation
ALOHA	=	Add-on Lantus® to Oral Hypoglycemic Agents
BLOOM-DM	=	Behavioral modification and Lorcaserin for Obesity and Overweight Management in Diabetes Mellitus
BLOSSOM	=	Behavioral modification and Lorcaserin Second Study for Obesity Management
EASIE	=	Evaluation of Insulin Glargine *versus* Sitagliptin in Insulin-Naïve Patients
EQUIP	=	Endpoint Quality and Intervention Program
FORTRESS	=	Fetal Outcome Retrospective Topiramate Exposure Study
IDA	=	Insulin Diabetes Angioplasty
LEAD-6	=	Liraglutide Effect and Action in Diabetes-6
NAVIGATOR Study	=	Nateglinide and Valsartan in Impaired Glucose Tolerance Outcomes Research ORIGIN: Outcome Reduction with an Initial Glargine Intervention
ORIGIN	=	Outcome Reduction with an Initial Glargine Intervention

PROactive   = Prospective Pioglit Azone Clinical Trial In macro Vascular Events

TSA   = Trial Sequential Analyses

UGDP   = University Group Diabetes Program

UKPDS   = United Kingdom Prospective Diabetes Study

VADT   = Veterans Affairs Diabetes Trial

REFERENCES

[1]   WHO Pharmaceuticals Newsletter 1999, No. 01&02. Available at: http: //apps. who.int/medicinedocs/en/d/ Js2268e/ 2.html#Js2268e.2.1 [accessed 1999].

[2]   Hu S, Wang S, Dunning BE. Glucose-dependent and glucose-sensitizing insulinotropic effect of nateglinide: Comparison to sulfonylureas and repaglinide. Int J Exp Diabetes Res 2001; 2: 63-72.

[3]   Leclercq-Meyer V, Ladriere L, Fuhlendorff J, Malaisse WJ. Stimulation of insulin and somatostatin release by two meglitinide analogs. Endocrine 1997; 7: 311-7.

[4]   Hollander PA, Schwartz SL, Gatlin MR, *et al*. Importance of early insulin secretion: comparison of nateglinide and glyburide in previously diet-treated patients with type 2 diabetes. Diabetes Care 2001; 24: 983-8.

[5]   Horton ES, Clinkingbeard C, Gatlin M, Foley J, Mallows S, Shen S. Nateglinide alone and in combination with metformin improves glycemic control by reducing mealtime glucose levels in type 2 diabetes. Diabetes Care 2000; 23: 1660-1665.

[6]   Saloranta C, Hershon K, Ball M, Dickinson S, Holmes D. Efficacy and safety of nateglinide in type 2 diabetic patients with modest fasting hyperglycemia. J Clin Endocr Metab 2002; 87: 4171-6.

[7]   Gromada J, Dissing S, Kofod H, Frokjaer-Jensen J. Effects of the hypoglycemic drugs repaglinide and glibenclamide on ATP-sensitive potassium-channels and cytosolic calcium levels in/3 TC3 cells and rat pancreatic beta cells. Diabetologia 1995; 38: 1025-32.

[8]   Malaisse WJ. Stimulation of insulin release by non-sulfonylurea hypoglycemic agents: the meglitiide family. Horm Metab Res 1995; 27: 263-6.

[9]   Fuhlendorff J, Rorsman P, Kofod H, *et al*. Stimulation of insulin release by repaglinide and glibenclamide involves both common and distinct processes. Diabetes 1998; 47: 345-51.

[10]   Hansen AMK, Hansen JB, Carr RD, Ashcroft FM, Wahl P. Kir6.2-dependent high-affinity repaglinide binding to β-cell K_{ATP} channels. Brit J Pharmacol 2005; 144: 551-7.

[11]   Reimann F, Proks P, Ashcroft FM. Effects of mitiglinide (S 21403) on Kir6.2/SUR1, Kir6.2/SUR2A and Kir6.2 /SUR2B types of ATP-sensitive potassium channel. Br J Pharmacol 2001; 132(7): 1542-8.

[12]   Toyoda Y, Mizutani K, Miwa I. Increase in hepatic glucose metabolism by mitiglinide calcium dehydrate-induced translocation of glucokinase from the nuclear to the cytoplasm. Prog Med 2008; 28: 1951-8.

[13] Gangji AS, Cuckierman T, Gerstein HC, Goldsmith CH, Clase CM. A systematic review and meta-analysis of hypoglycemia and cardiovascular events: a comparison of glyburide with other secretagogues and with insulin. Diabetes Care 2007; 30. 389-94.

[14] Tahrani AA, Piya MK, Barnett AH. Saxagliptin: a new DPP-4 inhibitor for the treatment of type 2 diabetes mellitus. Adv Ther 2009; 26(3): 249-62.

[15] Gautier JF, Choukem SP, Girard J. Physiology of incretins (GIP and GLP-1) and abnormalities in type 2 diabetes. Diabetes Metab 2008; 34 Suppl 2: S65-S72.

[16] Drucker DJ. The biology of incretin hormones. Cell Metab 2006; 3(3): 153-65.

[17] Barnett A. DPP-4 inhibitors and their potential role in the management of type 2 diabetes. Int J Clin Pract 2006; 60 (11): 1454-70.

[18] Kulasa KM, Henry RR. Pharmacotherapy of hyperglycemia. Expert Opin Pharmacother 2009; 10 (15): 2415-32.

[19] Kirby M, Ming Tse Yu D, O'Connor S, Gorrell M. Inhibitor selectivity in the clinical application of dipeptidyl peptidase-4 inhibition. Clin Sci 2010;118(1): 31-41.

[20] Henry R, Smith S, Schwartz S, et al. Effects of saxagliptin on β-cell stimulation and insulin secretion in patients with type 2 diabetes. Diabetes Obes Metab 2011; 13(9): 850-8.

[21] Mari A, Sallas WM, He YL, et al. Vildagliptin, a dipeptidyl peptidase-IV inhibitor, improves model-assessed beta-cell function in patients with type 2 diabetes. J Clin Endocr Metab 2005; 90(8): 4888-94.

[22] Berhan A, Berhan Y. Efficacy of alogliptin in type 2 diabetes treatment: a meta-analysis of randomized double-blind controlled studies BMC Endocr Disord 2013; 13: 9.

[23] D'Alessio DA, Denney AM, Hermiller LM, et al. Treatment with the dipeptidyl peptidase-4 inhibitor vildagliptin improves fasting islet-cell function in subjects with type 2 diabetes. J Clin Endocr Metab 2009; 94: 81-8.

[24] Edwards CM, Stanley SA, Davis R, et al. Exendin-4 reduces fasting and postprandial glucose and decreases energy intake in healthy volunteers. Am J Physiol Endocrinol Metab 2001; 281: E155-E61.

[25] Nielsen LL, Young AA, Parkes DG. Pharmacology of exenatide (synthetic exendin-4): a potential therapeutic for improved glycemic control of type 2 diabetes. Reg Pept 2004; 117: 77-88.

[26] Drucker DJ, Buse JB, Taylor K, et al. Exenatide once weekly *versus* twice daily for the treatment of type 2 diabetes: a randomised, open-label, non-inferiority study. Lancet 2008; 372: 1240-50.

[27] Kim D, MacConell L, Zhuang D, et al. Effects of once-weekly dosing of a long acting release formulation of exenatide on glucose control and body weight in subjects with type 2 diabetes. Diabetes Care 2007; 30: 1487-93.

[28] Ratner RE, Rosenstock J, Boka G DRI6012 Study Investigators. Dose-dependent effects of the once-daily GLP-1 receptor agonist lixisenatide in patients with type 2 diabetes inadequately controlled with metformin: a randomized, double-blind, placebo-controlled trial. Diabet Med 2010; 27: 1024-32.

[29] Werner U, Haschke G, Herling AW, Kramer W. Pharmacological profile of lixisenatide: A new GLP-1 receptor agonist for the treatment of type 2 diabetes. Reg Pept 2010; 164: 58-64.

[30] Buse JB, Rosenstock J, Sesti G, et al. Liraglutide once a day *versus* exenatide twice a day for type 2 diabetes: a 26-week randomised parallel-group, multinational, open-label trial (LEAD-6). Lancet 2009; 374(9683): 39-47.

[31] Amori RE, Lau J, Pittas AG. Efficacy and safety of incretin therapy in type 2 diabetes: systematic review and meta-analysis. J Am Med Assoc 200; 298 (2): 194-206.

[32] Koehler JA, Baggio LL, Lamont BJ, Ali S, Drucker DJ. Glucagon-like peptide-1 receptor activation modulates pancreatitis-associated gene expression but does not modify the susceptibility to experimental pancreatitis in mice. Diabetes 2009; 58 (9): 2148-61.

[33] Nachnani JS, Bulchandani DG, Nookala A, *et al.* Biochemical and histological effects of exendin-4 (exenatide) on the rat pancreas. Diabetologia. 2010; 53 (1): 153-9.

[34] Tatarkiewicz K, Smith PA, Sablan EJ, *et al.* Exenatide does not evoke pancreatitis and attenuates chemically induced pancreatitis in normal and diabetic rodents. Am J Physiol Endocrinol Metab 2010; 299 (6): E1076-86.

[35] Rios-Vazquez R, Marzoa-Rivas R, Gil-Ortega I, Kaski JC. Peroxisome proliferator-activated receptor-gamma agonists for management and prevention of vascular disease in patients with and without diabetes mellitus. Am J Cardiovasc Drugs 2006; 6: 231-42.

[36] Tontonoz P, Hu E, Graves RA, Budavari AI, Spiegelman BM. mPPAR gamma 2: tissue-specific regulator of an adipocyte enhancer. Genes Dev 1994; 8: 1224-34.

[37] Nedergaard J, Petrovic N, Lindgren EM, Jacobsson A, Cannon B. PPARgamma in the control of brown adipocyte differentiation. Biochim Biophys Acta 2005; 1740: 293-304.

[38] Dubois M, Pattou F, Kerr-Conte J, *et al.* Expression of peroxisome proliferatoractivated receptor gamma (PPAR gamma) in normal human pancreatic islet cells. Diabetologia 2000; 43: 1165-9.

[39] Willson TM, Lambert MH, Kliewer SA. Peroxisome proliferator-activated receptor gamma and metabolic disease. Annu Rev Biochem 2001; 70: 341-67.

[40] Norris AW, Chen L, Fisher SJ, *et al.* Muscle-specific PPARgamma-deficient mice develop increased adiposity and insulin resistance but respond to thiazolidinediones. J Clin Invest 2003; 112: 608-18.

[41] Boettcher E, Csako G, Pucino F, Wesley R, Loomba R. Meta-analysis: pioglitazone improves liver histology and fibrosis in patients with non-alcoholic steatohepatitis. Aliment Pharmacol Ther 2012; 35(1): 66-75.

[42] Home PD, Pocock SJ, Beck-Nielsen H, *et al.* Rosiglitazone evaluated for cardiovascular outcomes in oral agent combination therapy for type 2 diabetes (RECORD): a multicentre, randomised, open-label trial. Lancet 2009; 373(9681): 2125-35.

[43] Nesto RW, Bell D, Bonow RO, *et al.* Thiazolidinedione use, fluid retention, and congestive heart failure: a consensus statement from the American Heart Association and American Diabetes Association. Diabetes Care. 2004; 27(1): 256-63.

[44] Douglas IJ, Evans SJ, Pocock S, Smeeth L. The risk of fractures associated with thiazolidinediones: a self-controlled case-series study. PLOS Med 2009; 6(9): e1000154.

[45] Akune T, Ohba S, Kamekura S, *et al.* PPARgamma insufficiency enhances osteogenesis through osteoblast formation from bone marrow progenitors. J Clin Invest 2004; 113(6): 846-55.

[46] Isabelle N, Colmers BScH, Samantha L, *et al.* Use of thiazolidinediones and the risk of bladder cancer among people with type 2 diabetes: a meta-analysis. Can Med Assoc J 2012; 184 (12): E675-83.

[47] Kirpichnikov D, McFarlane SI, Sowers JR. Metformin: an update. Ann Intern Med 2002; 137: 25-33.

[48] Zhou G, Myers R, Li Y, *et al.* Role of AMP-activated protein kinase in mechanism of metformin action. J Clin Invest 2001; 108(8): 1167-74.

[49] Garber AJ, Duncan TG, Goodman AM, Mills DJ, Rohlf JL. Efficacy of metformin in type II diabetes: results of a double-blind, placebo-controlled, dose response trial. Am J Med 1997; 103(6): 491-7.

[50] Hoffmann J, Spengler M. Efficacy of 24-week monotherapy with acarbose, metformin, or placebo in dietary-treated NIDDM patients: the Essen-II Study. Am J Med 1997; 103: 483-90.

[51] UK Prospective Diabetes Study (UKPDS) Group. Effect of intensive blood glucose control with metformin on complications in overweight patients with type 2 diabetes (UKPDS 34). Lancet 1998; 352: 854-865.

[52] Tessier D, Maheux P, Khalil A, Fulop T. Effects of gliclazide *versus* metformin on the clinical profile and lipid peroxidation markers in type 2 diabetes. Metab Clin Exp 1999; 48: 897-903.

[53] Abbasi F, Kamath V, Rizvi AA, Carantoni M, Chen YD, Reaven GM. Results of a placebo-controlled study of the metabolic effects of the addition of metformin to sulfonylurea-treated patients. Evidence for a central role of adipose tissue. Diabetes Care 1997; 20(12): 1863-9.

[54] Landin K, Tengborn L, Smith U. Treating insulin resistance in hypertension with metformin reduces both blood pressure and metabolic risk factors. J Intern Med 1991; 229: 181-7.

[55] Inzucchi SE. Oral antihyperglycemic therapy for type 2 diabetes. J Am Med Assoc 2002; 287: 360-72.

[56] Morin-Papunen L, Rautio K, Ruokonen A, Hedberg P, Puukka M, Tapanainen JS. Metformin reduces servum C- reactive protein levels in women with polycystic ovary syndrome. J Clin Endocr Metab 2003; 88(10): 4649-54.

[57] Canadian Diabetes Association Clinical Practice Guidelines Expert Committee. Canadian Diabetes Association 2003 Clinical Practice Guidelines for the Prevention and Management of Diabetes in Canada. Can J Diabetes 2003; 27 (suppl 2): S1-S152.

[58] Ikeda T, Iwata K, Murakami H. Inhibitory effect of metformin on intestinal glucose absorption in perfused rat intestine. Biochem Pharmacol 2000; 59: 887-90.

[59] DeFronzo RA, Barzilai N, Simonson DC. Mechanism of metformin action in obese and lean noninsulin-dependent diabetic mellitus. N Engl J Med 1995; 333: 550-4.

[60] Salpeter S, Greyber E, Pasternak G, Salpeter E. Risk of fatal and nonfatal lactic acidosis with metformin use in type 2 diabetes mellitus. [Cochrane review]. In: The Cochrane Library; Issue 4, 2004. Oxford: Update Software.

[61] Currie CJ, Peters JR, Evans M. Dispensing patterns and financial costs of glucose-lowering therapies in the UK from 2000 to 2008. Diabetes Med 2010; 27: 744-52.

[62] American Diabetes Association. Standards of medical care in diabetes. Diabetes Care 2011; 34: S11-S61.

[63] Rensing KL, Reuwer AQ, Arsenault BJ, *et al*. Reducing cardiovascular disease risk in patients with type 2 diabetes and concomitant macrovascular disease: can insulin be too much of a good thing? Diabetes Obes Metab 2011; 13: 1073-87.

[64] Nandish S, Bailon O, Wyatt J, *et al*. Vasculotoxic effects of insulin and its role in atherosclerosis: what is the evidence? Curr Atheroscler Rep 2011; 13: 123-8.

[65] Lebovitz HE. Insulin: potential negative consequences of early routine use in persons with type 2 diabetes. Diabetes Care 2011; 34(suppl 2): s225-s30.

[66]   Currie CJ, Johnson JA. The safety profile of exogenous insulin in people with type 2 diabetes: justification for concern. Diabetes Obes Metab 2012; 14: 1-4.

[67]   Currie CJ, Poole CD, Evans M, Peters JR, Morgan CL. Mortality and other important diabetes-related outcomes with insulin *vs* other antihyperglycemic therapies in type 2 diabetes J Clin Endocr Metab 2013; 98: 668-77.

[68]   Becker RH. Insulin glulisine complementing basal insulins: a review of structure and activity. Diabetes Technol Ther 2007; 9: 109-21.

[69]   Koivisto VA. The human insulin analogue insulin lispro. Ann Med 1998; 30: 260-6.

[70]   DiMarchi RD, Chance RE, Long HB, Shields JE, Slieker LJ. Preparation of an insulin with improved pharmacokinetics relative to human insulin through consideration of structural homology with insulin-like growth factor I. Horm Res 1994; 41: 93-6.

[71]   Campbell RK, Campbell LK, White JR. Insulin lispro: its role in the treatment of diabetes mellitus. Ann Pharmacother 1996; 30: 1263-71.

[72]   Becker RH, Frick AD. Clinical pharmacokinetics and pharmacodynamics of insulin glulisine. Clin Pharmacokinet 2008; 47: 7-20.

[73]   Becker RH, Frick AD, Burger F, Potgieter JH, Scholtz H: Insulin glulisine, a new rapid-acting insulin analogue, displays a rapid time-action profile in obese non-diabetic subjects. Exp Clin Endocrinol Diabetes 2005; 113: 435-43.

[74]   Becker RH. Insulin glulisine complementing basal insulins: a review of structure and activity. Diabetes Technol Ther 2007; 9: 109-21.

[75]   Heise T, Nosek L, Spitzer H, *et al*. Insulin glulisine: a faster onset of action compared with insulin lispro. Diabetes Obes Metab 2007; 9: 746-53.

[76]   Thisted H, Johnsen SP, Rungby J. An update on the long-acting insulin analogue glargine. Basic Clin Pharmacol Toxicol 2006; 99: 1-11.

[77]   Kadowaki T, Ohtani T, Odawara M. Potential Formula for the Calculation of Starting and Incremental Insulin Glargine Doses: ALOHA Subanalysis PLOS One 2012; 7(8): e41358.

[78]   Chapman TM, Perry CM. Insulin detemir: a review of its use in the management of type 1 and 2 diabetes mellitus. Drugs 2004; 64(22): 2577-95.

[79]   Philips JC, Scheen A. Insulin detemir in the treatment of type 1 and type 2 diabetes. Vasc Health Risk Manag 2006; 2(3): 277-83.

[80]   Havelund S, Plum A, Ribel U, *et al*. The mechanism of protraction of insulin detemir, a long-acting, acylated analog of human insulin. Pharm Res 2004; 21(8): 1498-504.

[81]   Jonassen I, Havelund S, Ribel U, *et al*. Insulin degludec is a new generation ultra-long acting basal insulin with a unique mechanism of protraction based on multi-hexamer formation. Diabetes 2010; 59: A11.

[82]   Heise T, Hermanski L, Nosek L, Feldman A, Rasmussen S, Haahr H. Insulin degludec: four times lower pharmacodynamic variability than insulin glargine under steady-state conditions in type 1 diabetes. Diabetes Obes Metab 2012; 14(9): 859-64.

[83]   Freemantle N, Meneghini L, Christensen T, Wolden M L, Jendle J, Ratner R. Insulin degludec improves health- related quality of life (SF-36®) compared with insulin glargine in people with Type 2 diabetes starting on basal insulin: a meta-analysis of phase 3a trials. Diabet Med 2013; 30(2): 226-32.

[84]   Clinical guidelines on the identification, evaluation, and treatment of overweight and obesity in adults—the evidence report. Nat Inst Health Obes Res 1998; 6 (suppl 2): 51S-209S.

[85]   Guerciolini R. Mode of action of orlistat. Int J Obes Relat Metab Disord 1997; 21(Suppl 3): S12-23.

[86]   Sjöström L, Rissanen A, Andersen T, *et al*. Randomised placebo-controlled trial of orlistat for weight loss and prevention of weight regain in obese patients. Lancet 1998; 352: 167-72.

[87]   Davidson MH, Hauptman J, DiGirolamo M, *et al*. Weight control and risk factor reduction in obese subjects treated for 2 years with orlistat: a randomized controlled trial. J Am Med Assoc 1999; 281: 235-42.

[88]   Morris M, Lane P, Lee K, Parks D. An integrated analysis of liver safety data from orlistat clinical trials. Obes Facts 2012; 5: 485-94.

[89]   Douglas IJ, Langham J, Bhaskaran K, Brauer R, Smeeth L. Orlistat and the risk of acute liver injury: self controlled case series study in UK Clinical Practice Research Datalink Brit Med J 2013; 346: f1936.

[90]   Chawla S, Kaushik N, Singh NP, Ghosh RK, Saxena A. Effect of addition of either sitagliptin or pioglitazone in patients with uncontrolled type 2 diabetes mellitus on metformin: A randomized controlled trial. J Pharmacol Pharmacother 2013; 4(1): 27-32.

[91]   Pratley RE, Nauck M, Bailey T, *et al*. Liraglutide *versus* sitagliptin for patients with type 2 diabetes who did not have adequate glycaemic control with metformin: a 26-week, randomised, parallel-group, open-label trial. Lancet 2010; 375: 1447-56.

[92]   El-Ouaghlidi A, Rehring E, Holst JJ, *et al*. The dipeptidyl peptidase 4 inhibitor vildagliptin does not accentuate glibenclamide-induced hypoglycemia but reduces glucose-induced glucagon-like peptide 1 and gastric inhibitory polypeptide secretion. J Clin Endocr Metab 2007; 92 (11): 4165-71.

[93]   Buse JB, Rosenstock J, Sesti G, *et al*. Liraglutide once a day *versus* exenatide twice a day for type 2 diabetes: a 26-week randomised, parallel-group, multinational, open-label trial (LEAD-6). Lancet 2009; 374: 39-47.

[94]   Aschner P, Chan J, Owens DR, *et al*. Insulin glargine *versus* sitagliptin in insulin-naive patients with type 2 diabetes mellitus uncontrolled on metformin (EASIE): a multicentre, randomised open-label trial. Lancet 2012; 379 (9833): 262-2269.

[95]   Russell-Jones D, Vaag A, Schmitz O, *et al*. Liraglutide *vs* insulin glargine and placebo in combination with metformin and sulfonylurea therapy in type 2 diabetes mellitus (LEAD-5 met+SU): a randomised controlled trial. Diabetologia 2009; 52 (10): 2046-55

[96]   Ray KK, Seshasai SR, Wijesuriya S, *et al*. Effect of intensive control of glucose on cardiovascular outcomes and death in patients with diabetes mellitus: a meta-analysis of randomised controlled trials. Lancet 2009; 373: 1765-72.

[97]   Holman RR, Paul SK, Bethal MA, Matthews DR, Neil HA. 10-year follow-up of intensive glucose control in type 2 diabetes. N Engl J Med 2008; 359: 1577-89.

[98]   Charbonnel B, Dormandy J, Erdmann E, Massi-Benedetti M, Skene A. The prospective pioglitazone clinical trial in macrovascular events (PROactive): can pioglitazone reduce cardiovascular events in diabetes? Study design and baseline characteristics of 5238 patients. Diabetes Care. 2004; 27(7): 1647-53.

[99]   Dormandy JA, Charbonnel B, Eckland DJ, *et al*. Secondary prevention of macrovascular events in patients with type 2 diabetes in the PROactive Study (PROspective pioglitAzone Clinical Trial In macroVascular Events): a randomised controlled trial. Lancet 2005; 366 (9493): 1279-89.

[100] Wilcox R, Kupfer S, Erdmann E. Effects of pioglitazone on major adverse cardiovascular events in high-risk patients with type 2 diabetes: results from PROspective pioglitAzone Clinical Trial In macro Vascular Events (PROactive 10). Am Heart J 2008; 155(4): 712-7.

[101] The ADVANCE Collaborative group. Intensive blood glucose control and vascular outcomes in patients with type 2 diabetes. New Engl J Med 2008; 358 (24): 2560-72.

[102] The action to control cardiovascular risk in diabetes study group Effect of intensive glucose lowering in type 2 diabetes. New Engl J Med 2008; 358 (24): 2545-59.

[103] Duckworth W, Abraira C, Moritz T, *et al.* Glucose control and vascular complications in veterans with type 2 diabetes. New Engl J Med 2009; 360: 129-39.

[104] Cholesterol Treatment Trialists' (CTT) Collaborators. Efficacy of cholesterol-lowering therapy in 18,686 people with diabetes in 14 randomised trials of statins: a meta-analysis. Lancet 2008; 371(9607): 117-25.

[105] UK Prospective Diabetes Study Group. Tight blood pressure control and risk of macrovascular and microvascular complications in type 2 diabetes: UKPDS 38. Brit Med J 1998; 317(7160): 703-13.

[106] Hansson L, Zanchetti A, Carruthers SG, *et al.* Effects of intensive blood-pressure lowering and low-dose aspirin in patients with hypertension: principal results of the Hypertension Optimal Treatment (HOT) randomised trial. HOT Study Group. Lancet 1998; 351(9118): 1755-62.

[107] Turnbull F, Neal B, Algert C, *et al.* Effects of different blood pressure-lowering regimens on major cardiovascular events in individuals with and without diabetes mellitus: results of prospectively designed overviews of randomized trials. Arch Intern Med 2005; 165(12): 1410-9.

[108] Hemmingsen B, Lund SS, Gluud C, *et al.* Intensive glycemic control for patients with type 2 diabetes: systematic review with meta-analysis and trial sequential analysis of randomized clinical trials. Brit Med J 2011; 343: d4169.

[109] Thorlund K, Imberger G, Walsh M, *et al.* The number of patients and events required to limit the risk of overestimation of intervention effects in meta-analysis--a simulation study. PLOS One. 2011; 6(10): e25491.

[110] Wetterslev J, Thorlund K, Brok J, Gluud C. Trial sequential analysis may establish when firm evidence is reached in cumulative meta-analysis. J Clin Epidemiol. 2008; 61(1): 64-75.

[111] Hage C, Norhammar A, Grip L, *et al.* Glycaemic control and restenosis after percutaneous coronary interventions in patients with diabetes mellitus: a report from the Insulin Diabetes Angioplasty study. Diab Vasc Dis Res 2009; 6(2): 71-9.

[112] Yngen M, Norhammar A, Hjemdahl P, Wallén NH. Effects of improved metabolic control on platelet reactivity in patients with type 2 diabetes mellitus following coronary angioplasty. Diab Vasc Dis Res 2006; 3(1): 52-6.

[113] Higgins JPT, Green S. Cochrane handbook for systematic reviews of intervention 5.0.0. Cochrane Collaboration 2008 (available from Cochrane-handbook.org).

[114] Gallwitz B, Rosenstock J, Rauch T, *et al.* 2-year efficacy and safety of linagliptin compared with glimepiride in patients with type 2 diabetes inadequately controlled on metformin: a randomized, double-blind, non-inferiority trial. Lancet 2012; 380: 475-83.

[115] NAVIGATOR Study Group, Holman RR, Haffner SM, *et al.* Effect of nateglinide on the incidence of diabetes and cardiovascular events. New Engl J Med 2010; 362(16):1463-76.

[116] The ORIGIN Trial Investigators, Basal insulin and cardiovascular and other outcomes in dysglycemia. New Engl J Med 2012; 367(4): 319-28.

[117] Currie CJ, Poole CD, Evans M, Peters JR, Morgan CL. Mortality and other important diabetes-related outcomes with insulin *vs* other antihyperglycemic therapies in type 2 diabetes. J Clin Endocr Metab 2013; 98(2): 668-77.

[118] Santer R, Kinner M, Lassen CL, *et al*. Molecular analysis of the SGLT2 gene in patients with renal glucosuria. J Am Soc Nephrol 2003; 14: 2873-82.

[119] Hanefeld M, Forst T. Dapagliflozin, an SGLT2 inhibitor, for diabetes. Lancet 2010; 375: 2196-8.

[120] Clar C, Gill JA, Court R, Waugh N. Systematic review of SGLT2 receptor inhibitors in dual or triple therapy in type 2 diabetes. Brit Med J Open 2012; 2: e001007.

[121] Burki TK. FDA rejects novel diabetes drug over safety fears. Lancet 2012; 379: 507.

[122] Kaku K, Watada H, Iwamoto Y, *et al*. Efficacy and safety of monotherapy with the novel sodium/glucose cotransporter-2 inhibitor tofogliflozin in Japanese patients with type 2 diabetes mellitus: a combined Phase 2 and 3 randomized, placebo-controlled, double-blind, parallel-group comparative study. Cardiovasc Diabetol 2014; 13: 65

[123] Fyle M, White J, Taylor A, *et al*. Glucokinase activator PSN-GK1 displays enhanced antihyperglycemic and insulinotropic actions. Diabetologia 2007; 50: 1277-87.

[124] Pal M. Recent advances in glucokinase activators for the treatment of type 2 diabetes. Drug Discov Taday 2009; 14: 784-92.

[125] Matschinsky FM. Assessing the potential of glucokinase activators in diabetes therapy. Nat Rev Drug Discov 2009; 8: 399-416.

[126] O'Neil P, Smith SR, Weissman NJ, *et al*. Randomized placebo-controlled clinical trial of lorcaserin for weight loss in type 2 diabetes mellitus: the BLOOM-DM study. Obesity (Silver Spring, Md) 2012; 20: 1426-36.

[127] Fidler MC, Sanchez M, Raether B, *et al*. BLOSSOM clinical trial group. A one-year randomized trial of lorcaserin for weight loss in obese and overweight adults: the BLOSSOM trial. J Clin Endocr Metab 2011; 96: 3067-77.

[128] Smith SR, Weissman NJ, Anderson CM, *et al*. Behavioral Modification and Lorcaserin for Overweight and Obesity Management (BLOOM) Study Group. Multicenter, placebo-controlled trial of lorcaserin for weight management. N Engl J Med 2010; 363(3): 245-56.

[129] Lexicomp. http: //online.lexi.com.ezproxy.xula.edu/lco/action/home. Lexicomp Online Web Site. Accessed Nov 2012.

[130] Bray GA. Drug insight: appetite suppressants. Nat Clin Pract Gastr 2005; 2(2): 89-95.

[131] Adipex-P [package insert]. Sellersville, PA Teva Pharmaceuticals, USA 2012.

[132] Li Z, Maglione M, Tu W, *et al*. Meta-analysis: pharmacologic treatment of obesity. Ann Intern Med 2005; 142 (7): 532-46.

[133] Bray GA, Hollander P, Klein S, *et al*. A 6-month randomized, placebo-controlled, dose-ranging trial of topiramate for weight loss in obesity. Obes Res 2003; 11(6): 722-33.

[134] Bays HE, Gadde KM. Phentermine/topiramate for weight reduction and treatment of adverse metabolic consequences in obesity. Drugs Today (Barc) 2011; 47(12): 903-14.

[135] Bays H. Phentermine, topiramate and their combination for the treatment of adiposopathy ('sick fat') and metabolic disease. Expert Rev Cardiovasc Ther 2010; 8(12): 1777-801.

[136] Cameron F, Whiteside G, McKeage K. Phentermine and topiramate extended release (qsymia): first global approval. Drugs 2012; 72(15): 2033-42.

[137]   Malgarini RB, Pimpinella G. Phentermine plus topiramate in the treatment of obesity. Lancet 2011; 378(9786): 125-6; author reply 126-7.

[138]   Mercer SL. ACS chemical neuroscience molecule spotlight on Qnexa. ACS Chem Neurosci 2011; 2(4): 183-4.

[139]   Shah K, Villareal DT. Combination treatment to CONQUER obesity? Lancet 2011; 377(9774): 1295-7.

[140]   Garvey WT, Ryan DH, Look M, *et al.* Two-year sustained weight loss and metabolic benefits with controlled-release phentermine/topiramate in obese and overweight adults (SEQUEL): a randomized, placebo-controlled, phase 3 extension study. Am J Clin Nutr 2012; 95(2): 297-308.

[141]   Allison DB, Gadde KM, Garvey WT, *et al.* Controlled-release phentermine/topiramate in severely obese adults: a randomized controlled trial (EQUIP). Obesity (Silver Spring) 2012; 20(2): 330-42.

[142]   Aronne LJ, Peterson C, Troupin B, *et al.* Weight loss with V1-0521 (phentermine/controlled release topiramate) stops progression towards type 2 diabetes in obese non-diabetic subjects. Poster PO.22 presented at: 2010 Obesity Society meeting and the abstract included in Obesity Facts.

[143]   Gadde KM, Allison DB, Ryan DH, *et al.* Effects of low-dose, controlled-release, phentermine plus topiramate combination on weight and associated comorbidities in overweight and obese adults (CONQUER): a randomised, placebo-controlled, phase 3 trial. Lancet. 2011; 377(9774): 1341-52.

[144]   VIVUS I. VIVUS Reports Topline Findings from FORTRESS. [media release on the Internet].   December   21,   2011.   Available   at   http: //ir.vivus.com/releasedetail.cfm?ReleaseID=634920.

[145]   Chhabra P, Brayman KL. *Review Article* Current Status of Immunomodulatory and Cellular Therapies in Preclinical and Clinical Islet Transplantation Journal of Transplantation 2011; 2011, Article ID 637692, 24 pages

[146]   Choi JH, Banks AS, Estall JL, *et al.* Anti-diabetic drugs inhibit obesity-linked phosphorylation of PPARγ by Cdk5 Nature 2010; 466: 451-6.

[147]   Choi JH, Banks AS, Kamenecka TM, *et al.* Antidiabetic actions of a non- agonist PPARγ ligand blocking Cdk5-mediated phosphorylation Nature 2011; 477(7365): 477-81.

[148]   Minokoshi Y, Kim YB, Peroni OD, *et al.* Leptin stimulates fatty-acid oxidation by activating AMP-activated protein kinase. Nature 2002; 415: 339-43.

[149]   Minokoshi Y, Toda C, Okamoto S. Regulatory role of leptin in glucose and lipid metabolism in skeletal muscle. Indian J Endocrinol Metab 2012; 16(Suppl 3): S562-S568.

[150]   Coppari R, Bjorbaek C. Leptin revisited: its mechanism of action and potential for treating diabetes. Nat Rev Drug Discov 2012;11: 692-708.

[151]   Moon HS, Dalamaga M, Kim SY, *et al.* Leptin's role in lipodystrophic and nonlipodystrophic insulin-resistant and diabetic individuals. Endocr Rev 2013;34: 377-412.

[152]   Yamauchi T, Nio Y, Maki T, *et al.* Targeted disruption of AdipoR1 and AdipoR2 causes abrogation of adiponectin binding and metabolic actions. Nat Med 2007; 13: 332-9.

[153]   Tsuchida A, Yamauchi T, Ito Y, *et al.* Insulin/Foxo1 pathway regulates expression levels of adiponectin receptors and adiponectin sensitivity. J Biol Chem 2004; 279: 30817-22.

[154]   Sun Y, Zang Z, Zhong L, *et al.* Identification of adiponectin receptor agonist utilizing a fluorescence polarization based high throughput assay PLOS ONE 2013; 8 (5): e63354

[155] Bordone L, Guarente L. Calorie restriction, SIRT1 and metabolism: understanding longevity. Nat Rev Mol Cell Bio 2005; 6: 298-305.

[156] Cohen HY, Miller C, Bitterman KJ, *et al.* Calorie restriction promotes mammalian cell survival by inducing the SIRT1 deacetylase. Science 2004; 305: 390-2.

[157] Heilbronn LK, Civitarese AE, Bogacka I, Smith SR, Hulver M, Ravussin E. Glucose tolerance and skeletal muscle gene expression in response to alternate day fasting. Obes Res 2005; 13: 574-81.

[158] Nisoli E, Tonello C, Cardile A, *et al.* Calorie restriction promotes mitochondrial biogenesis by inducing the expression of eNOS. Science 2005; 310: 314-7.

[159] Frye RA. Characterization of five human cDNAs with homology to the yeast SIR2 gene: Sir2-like proteins (sirtuins) metabolize NAD and may have protein ADP-ribosyltransferase activity. Biochem Biophys Res Commun 1999; 260: 273-9.

[160] Frye RA. Phylogenetic classification of prokaryotic and eukaryotic Sir2-like proteins. Biochem Biophys Res Commun 2000; 273: 793-8.

[161] Imai S, Armstrong CM, Kaeberlein M, Guarente L. Transcriptional silencing and longevity protein Sir2 is an NAD- dependent histone deacetylase. Nature 2000; 403: 795-800.

[162] Milne JC, Lambert PD, Schenk S, *et al.* Small molecule activators of SIRT1 as therapeutics for the treatment of type 2 diabetes. Nature 2007; 450(7170): 712-6.

[163] Kuo LE, Kitlinska JB, Tilan JU, *et al.* Neuropeptide Y acts directly in the periphery on fat tissue and mediates stress-induced obesity and metabolic syndrome. Nat Med 2007;13: 803-11.

[164] Schwartz MW, Porte Jr D, Seeley RJ, Baskin DG. Central nervous system control of food intake. Nature 2000; 404: 661-71.

[165] McMinn JE, Baskin DG, Schwarz MW. Neuroendocrine mechanisms regulating food intake and body weight. Obes Rev 2000;1: 37-46.

[166] Broberger C. Brain regulation of food intake and appetite: molecules and networks. J Intern Med 2005; 258: 301-27.

[167] Karvonen MK, Pesonen U, Koulu M, *et al.* Association of a leucine (7)-to- proline (7) polymorphism in the signal peptide of neuropeptide Y with high serum cholesterol and LDLcholesterol levels. Nat Med 1998; 4: 1434-7.

[168] Karvonen MK, Koulu M, Pesonen U, *et al.* Leucine 7 to proline 7 polymorphism in the preproneuropeptide Y is associated with birth weight and serum triglyceride concentration in preschool aged children. J Clin Endocr Metab 2000; 85: 1455-60.

[169] Niskanen L, Karvonen MK, Valve R, *et al.* Leucine 7 to proline 7 polymorphism in the neuropeptide Y gene is associated with enhanced carotid atherosclerosis in elderly patients with type 2 diabetes and control subjects. J Clin Endocr Metab 2000; 85: 2266-9.

[170] Hung CC, Pirie F, Luan J, *et al.* Studies of the peptide YY and neuropeptide Y2 receptor genes in relation to human obesity and obesity-related traits. Diabetes 2004; 53: 2461-6.

[171] Lavebratt C, Alpman A, Persson B, Arner P, Hoffstedt J. Common neuropeptide Y2 receptor gene variant is protective against obesity among Swedish men. Int J Obes 2006; 30: 453-9.

[172] Campbell CD, Lyon HN, Nemesh J, *et al.* Association studies of BMI and type 2 diabetes in the neuropeptide Y pathway: a possible role for NPY2R as a candidate gene for type 2 diabetes in men. Diabetes 2007; 56: 1460-7.

[173] Siddiq A, Gueorguiev M, Samson C, *et al*. Single nucleotide polymorphisms in the neuropeptide Y2 receptor (NPY2R) gene and association with severe obesity in French white subjects. Diabetologia 2007; 50: 574-84.

[174] Torekov SS, Larsen LH, Andersen G, *et al*. Variants in the 5′ region of the neuropeptide Y receptor Y2 gene (NPY2R) are associated with obesity in 5,971 white subjects. Diabetologia 2006; 49: 2653-8.

[175] Takiguchi E, Fukano C, Kimura Y, Tanaka M, Tanida K, Kaji H. Variation in the 5′-flanking region of the neuropeptide Y2 receptor gene and metabolic parameters. Metabolism 2010; 59: 1591-6.

[176] Morgan SA, Sherlock M, Gathercole LL, *et al*. 11β-Hydroxysteroid dehydrogenase type 1 regulates glucocorticoid-induced insulin resistance in skeletal muscle. Diabetes 2009; 58: 2506-15.

[177] Hawkins M, Hunter D, Kishore P, *et al*. INCB013739, a Selective inhibitor of 11β-hydroxysteroid dehydrogenase type 1 (11β-HSD1), improves insulin sensitivity and lower plasma cholesterol over 28 days in patients with type 2 diabetes mellitus (Abstract). 68th Scientific Sessions of the American Diabetes Association, 6-10 June 2008, Moscone Convention Center, SanFrancisco, California.

[178] Winnick JJ, Ramnanan CJ, Saraswathi V, *et al*. Effects of 11β-hydroxysteroid dehydrogenase-1 inhibition on hepatic glycogenolysis and gluconeogenesis. Am J Physiol Endocrinol Metab. 2013; 304(7): E747-56.

[179] Spranger J, Kroke A, Möhlig M, *et al*. Inflammatory cytokines and the risk to develop type 2 diabetes: results of the prospective population-based European Prospective Investigation into Cancer and Nutrition (EPIC)-Potsdam Study. Diabetes 2003; 52: 812-7.

[180] Herder C, Haastert B, Müller-Scholze S, *et al*. Association of systemic chemokine concentrations with impaired glucose tolerance and type 2 diabetes: results from the Cooperative Health Research in the Region of Augsburg Survey S4 (KORA S4). Diabetes 2005; 54 (suppl. 2): S11-7.

[181] Mohamed-Ali V, Goodrick S, Rawesh A, *et al*. Subcutaneous adipose tissue releases interleukin-6, but not tumor necrosis factor-α, *in vivo*. J Clin Endocr Metab 1997; 82: 4196-200.

[182] Ellingsgaard H, Hauselmann I, Schuler B, *et al*. Interleukin-6 enhances insulin secretion by increasing glucagon-like peptide-1 secretion from L cells and alpha cells. Nat Med 2011; 17 (11): 1481-90.

[183] Yi P, Park J-S, Melton DA. Betatrophin: a hormone that controls pancreatic β cell proliferation. Cell 2013; 153: 1- 12.

[184] Nagpal S, Patel S, Asano AT, Johnson AT, Duvic M, Chandraratna RA. Tazarotene-induced gene 2 (TIG2), a novel retinoid-responsive gene in skin. J Invest Dermatol 1997; 109: 91-5.

[185] Wittamer V, Franssen JD, Vulcano M, *et al*. Specific recruitment of antigen- presenting cells by chemerin, a novel processed ligand from human inflammatory fluids. J Exp Med 2003; 198: 977-85.

[186] Goralski KB, McCarthy TC, Hanniman EA, *et al*. Chemerin, a novel adipokine that regulates adipogenesis and adipocyte metabolism. J Biol Chem 2007; 282: 28175-88.

[187] Bozaoglu K, Bolton K, McMillan J, *et al*. Chemerin is a novel adipokine associated with obesity and metabolic syndrome. Endocrinology 2007; 148: 4687-94.

[188] Roh SG, Song SH, Choi KC, *et al*. Chemerin—a new adipokine that modulates adipogenesis *via* its own receptor. Biochem Biophys Res Commun 2007; 362: 1013-8.

[189] Takahashi M, Takahashi Y, Takahashi K, *et al*. Chemerin enhances insulin signaling and potentiates insulin stimulated glucose uptake in 3T3-L1 adipocytes. FEBS Lett 2008; 582: 573-8.

[190] Takahashi M, Okimura Y, Iguchi G, *et al*. Chemerin regulates β-cell function in mice. Scientific Reports 2011; 1: 123 | DOI: 10.1038/srep00123.

[191] Mataki C, Magnier BC, Houten SM, *et al*. Compromised intestinal lipid absorption in mice with a liver specific deficiency of the liver receptor homolog 1. Mol Cell Biol 2007; 27: 8330-9.

[192] Lee YK, Schmidt DR, Cummins CL, *et al*. Liver receptor homolog-1 regulates bile acid homeostasis but is not essential for feedback regulation of bile acid synthesis. Mol Endocrinol 2008; 22: 1345-56.

[193] Krylova IN, Sablin EP, Moore J, *et al*. Structural analyses reveal phosphatidyl inositols as ligands for the NR5 orphan receptors SF-1 and LRH-1. Cell 2005; 120: 343-55.

[194] Ortlund EA, Lee Y, Solomon IH, *et al*. Modulation of human nuclear receptor LRH-1 activity by phospholipids and SHP. Nature Struct Mol Biol 2005; 12: 357-63.

[195] Wang W, Zhang C, Marimuthu A, *et al*. The crystal structures of human steroidogenic factor-1 and liver receptor homologue-1. Proc Natl Acad Sci USA 2005; 102: 7505-10.

[196] Lee JM, Lee YK, Mamrosh JL, *et al*. A nuclear-receptor-dependent phosphatidylcholine pathway with antidiabetic effects. Nature 2011; 474: 506-12.

[197] Boyer S, Sharp PA, Debnam ES, Baldwin SA, Srai SKS. Streptozotocin diabetes and the expression of GLUT1 at the brush border and basolateral membranes of intestinal enterocytes. Febs Lett 1996; 396: 218-22.

[198] Wild GE, Turner R, Chao LGS, *et al*. Dietary lipid modulation of Na$^+$ glucose co-transporter (SGLT1), Na$^+$/K$^+$ ATPase, and ornithine decarboxylase gene expression in the rat small intestine in diabetes mellitus. Nutr Biochem 1997; 8: 673-80.

[199] Li JM, Che CT, Lau CBS, Leung PS, Cheng CHK. Inhibition of intestinal and renal Na$^+$-glucose cotransporter by naringenin. Int. J Biochem Cell B. 2006; 38: 985-95.

[200] Zheng Y, Scow JS, Duenes JA, Rochester SMG. Mechanisms of glucose uptake in intestinal cell lines: Role of GLUT2. Surgery 2012; 151: 13-25.

[201] Pari L, Srinivasan S. Antihyperglycemic effect of diosmin on hepatic key enzymes of carbohydrate metabolism in streptozotocin-nicotinamide-induced diabetic rats. Biomed Pharmacother 2010; 64: 477-81.

[202] Ugochukwu NH, Figgers CL. Modulation of the flux patterns in carbohydrate metabolism in the livers of streptozoticin-induced diabetic rats dietary caloric restriction. Pharmacol Res 2006; 54: 172-80.

[203] Li Y-G, Ji D-F, Zhong S, Lin T-B, Lv Z-Q, Hu G-Y. 1-deoxynojirimycin inhibits glucose absorption and accelerates glucose metabolism in streptozotocin-induced diabetic mice Scientific Reports 2013; 3: 1377 | DOI: 10.1038/srep01377

[204] Kharitonenkov A. Shiyanova TL, Koester A, *et al*. FGF-21 as a novel metabolic regulator. J Clin Invest 2005; 115: 1627-35.

[205] Dutchak PA, Katafuchi T, Bookout AL, *et al*. Fibroblast growth factor-21 regulates PPAR☐ activity and the antidiabetic actions of thiazolidinediones. Cell 2012; 148: 556-67.

[206] Reiman ML FGF21 mimetics shows therapeutic promise Cell Metab 2013; 18: 307-9

[207] Jonker JW, Suh JM, Atkins AR, *et al*. A PPARγ-FGF1 axis is required for adaptive adipose remodeling and metabolic homeostasis. Nature 2012; 485: 391-4.

[208] Sun, K and Scherer PE. The PPARγ-FGF1 axis: an unexpected mediator of adipose tissue homeostasis. Cell Res 2012; 22: 1416-1418.

[209] Suh JM, Jonker JW, Ahmadian M, *et al*. Endocrinization of FGF1 produces a neomorphic and potent insulin sensitizer. Nature 2014; 513: 436-9.

[210] Cohen P, Ntambi JM, Friedman JM. Stearoyl-CoA desaturase-1 and the metabolic syndrome. Curr Drug Targets Immune Endocr Metabol Disord 2003; 3: 271-80.

[211] Dobrzyn A, Ntambi JM. Stearoyl-CoA desaturase as a new drug target for obesity treatment. Obes Rev 2005; 6: 169-174.

CHAPTER 2

Antidiabetic Drugs of Type 2 Diabetes and Obesity

Tatjana Ábel[1,*] and Gabriella Lengyel[2]

[1]Outpatient Department, Military Hospital, Budapest Hungary; Faculty of Health Sciences, Semmelweis University, Budapest, Hungary and [2]2nd Department of Internal Medicine, Medical Faculty, Semmelweis University, Budapest, Hungary

Abstract: Overweight and obesity are risk factors for type 2 diabetes, and they also influence the overall prognosis. Almost 90 % of type 2 diabetic subjects are overweight and more than half are obese. Type 2 diabetes is a progressive metabolic disorder conferring an increased risk of cardiovascular comorbidities, including hypertension, coronary artery disease and stroke. The availability of several different classes of antidiabetic drugs in the last few years has increased the number of choices for physicians. The aim of these therapies is to reduce and maintain glucose concentrations and thereby prevent development of complications. When choosing between different antidiabetic drugs attention should be paid to their effect on body weight too.

Keywords: Adipose tissue, antidiabetic, BMI, bromocriptine, carbohydrate, colesevelam, complication, diabetes, DPP4 inhibitors; gastrointestinal, glinides, GLP-1 receptor agonists, HbA1c, hypoglycemia, hypothalamus, insulin, kidney, metabolism, metformin, obesity, pramlintide, SGLT2 inhibitors, sulfonylurea, thiazolidinedione.

INTRODUCTION

According to a survey of the World Health Organization (WHO), there was more than 1.9 billion overweight and 600 million obese adults worldwide in 2014 [1]. On average, 13 % of the adult population is obese (men 11 %, women 15 %), and 39 % is overweight (men 38 %, women 40 %). According to their calculation, the prevalence of obesity has doubled between 1980 and 2014.

In 2011, approximately 366 million diabetic patients were registered worldwide [2]. In Europe, 52.8 million people, aged between 20 and 79 years had diabetes, representing approx. 8.1 % of the population [3]. In 2014, however, diabetes

*Corresponding author Tatjana Ábel: Outpatient Department, Military Hospital, Budapest Hungary; Faculty of Health Sciences, Semmelweis University, Budapest, Hungary; Tel: +36-20-9-29-23-72; Fax: +36-1-397-48-55; E-mail: abelt@t-online.hu

affected already 9 % of the population according to the WHO [4]. Diabetes-related mortality affected 1.5 million people in 2012. Their estimations show that diabetes will be the 7th cause of death by 2030.

The risk of developing diabetes increases three times when the BMI is between 25.0 kg/m^2 and 29.9 kg/m^2, and twenty times if the BMI is 35 kg/m^2 or more, in comparison to those with a normal BMI value (18.5–24.9 kg/m^2) [5]. Therefore, the prevention or treatment of diabetes is of eminent importance, partly as it affects a great number of people, while obesity is often associated with a disturbed carbohydrate metabolism, which increases the risk of micro- and macrovascular complications in patients with diabetes. However, a systematic review of observational studies found that the risk for cardiovascular mortality and hospitalization was lowest in overweight patients compared with normal weight and obese patients (obesity paradox) [6]. This protective effect of overweight was not modified by age [7]. Nevertheless, the existence of the obesity paradox cannot lead to an underestimation of obesity as a crucial risk factor for the development of cardiovascular and metabolic diseases.

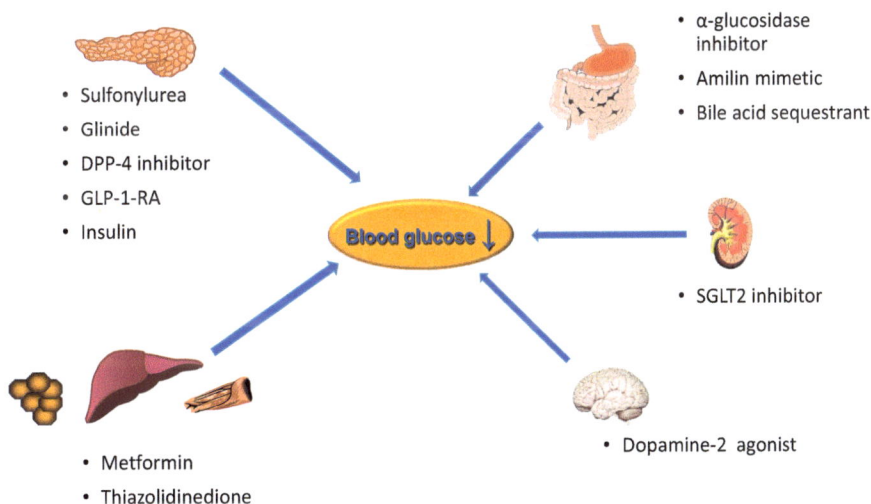

Figure 1: Antidiabetic agents used in the therapy of type 2 diabetes mellitus. DPP-4 = dipeptidyl peptidase 4; GLP-1-RA = glucagon-like peptide 1 receptor agonist; SGLT-2 = sodium-glucose co-transporter 2.

The significant spread of type 2 diabetes has contributed to the development of medicines with newer and newer points of attack (Fig. **1**). The importance of insulin resistance and ß-cell dysfunction to the pathogenesis of type 2 diabetes was debated for a long time. The availability of several different classes of drugs in the last few years that lower blood glucose and HbA1c has increased the number of choices for the physicians and patients too.

The therapy is aimed primarily at the control of carbohydrate metabolism. In addition to it, it is also important that the medicines used during the therapy should cause as few side effects, such as severe hypoglycemia, as possible, on the other hand they should also possess other beneficial effects, *e.g.* reduction of body weight. Ultimately, obesity in diabetic patients contributes to increased cardiovascular disease and microvascular complications too.

In this chapter, we summarized the effects exerted by medicines on body weight which are used in the therapy of type 2 diabetes mellitus.

METFORMIN

Metformin is a drug of first choice in the therapy of type 2 diabetic patients according to the recommendations of both the American Diabetes Association and the European Association for the Study of Diabetes [8, 9]. Metformin reduces hepatic glucose production, improves peripheral insulin sensitivity, and it inhibits gastrointestinal glucose absorption as well [10]. It reduces fasting blood glucose levels and the level of hemoglobin A1c (HbA1c) too (1.0-2.0 %), with a very low chance of any occurrence of an episode of hypoglycemia during its use [8]. Metformin can be recommended particularly for type 2 diabetic patients with overweight, as its use promotes a moderate reduction of body weight (0.6–2.7 kg), and also its preservation [11-13].

The mechanism of the body weight-reducing effects of metformin has not yet been exactly elucidated. Results published up to now (mostly from animal experiments) showed the decisive importance of the tissue-specific increase or decrease in the levels of the adenosine monophosphate-activated protein (AMPK) [14]. Upon the effect of metformin, AMPK levels decrease in the hypothalamus, and increase in the muscle, adipose tissue and in the liver.

Data of animal experiments showed that metformin can cross the blood-brain barrier, so that AMPK level decreases in the *hypothalamus* as a consequence [15].

Due to this, the intake of food decreases and the grade of insulin sensitivity increases eventually.

Metformin therapy can increase AMPK levels in the *muscle tissue*, therefore insulin-stimulated glucose uptake and fatty acid oxidation increase [16].

In addition to the reduction of carbohydrate absorption, metformin therapy also exerts other effects on the *gastrointestinal system* [17]. In part it increases the concentration of glucagon-like peptide-1 (GLP-1), which reduces gastric motility and emptying, and thus the amount of the energy taken in. Beyond this, animal experiments showed that metformin also influences intestinal flora (through its direct effect on enterocytes), and consequently it may regulate energy homeostasis as well [18]. Metformin may promote weight loss by activation of AMPK in the gut too [19]. Recently, demonstrated that metformin has a complex effect due to gut-based pharmacology in patients with type 2 diabetes [19].

Diabetic patients should be cautioned about gastrointestinal side effects (more common: abdominal discomfort, diarrhea) of metformin therapy, if these occur, the metformin dose may be lowered to a tolerated dose or exchange it for the long-acting formulation (extended-release).

Metformin increases AMPK level also in the *adipose tissue*, consequently leptin concentration and also the body weight may decrease [20].

Metformin therapy reduces the accumulation of ectopic fat, *e.g.*, in the *liver* [21]. Hepatic cholesterol and fatty acid synthesis and glucose production decrease during the therapy.

THIAZOLIDINEDIONES

Thiazolidinediones (TZDs) exert their blood glucose-lowering and HbA1c reducing (0.5-1.4 %) effect by increasing the activity of proliferator-activated receptor gamma (PPARγ). As a consequence of this, insulin-mediated glucose uptake increases primarily in the adipose tissue, but also in the liver and muscle tissue. In addition, they increase adiponectin production, subsequently insulin sensitivity [22].

An increase of body weight has been reported during TZD therapy (1.3–4.8 kg) [23]. This can partly be explained by the fact that, subcutaneous adipose tissue may increase. Nevertheless, it reduces the volume of visceral adipose tissue that is important in relation to decrease cardiovascular risk and improves insulin sensitivity. Note worthily, the waist-to-hip ratio will eventually decrease due to TZD therapy [24]. Another possible explanation of the growth of body weight observed during TZD treatment may be the fluid retention which may be generated during the therapy [25].

Of the TDZs, rosiglitazone has not been marketed in Europe. However, pioglitazone is an available choice for treating patients with type 2 diabetes mellitus.

DPP-4 INHIBITORS

The dipeptidyl peptidase-4 (DPP-4) inhibitors (sita-, saxa-, lina-, alogliptin and vildagliptin /not licensed in the U.S.) inhibit the degradation of glucagon-like peptide-1 (GLP-1) produced by the intestine. It improves fasting glucose control, reduces HbA1c by 0.5 to 0.8 %. Their effect on the body weight is neutral.

GLP-1 RECEPTOR AGONISTS

During the past two decades the physiology and pharmacology of GLP-1 and GLP-1 analogs in glucose, food intake, and body weight control have been gradually dissected. The GLP-1 receptor agonists (exenatide, exenatide extended release, liraglutide, albiglutide, dulaglutide and lixisenatide /not licensed in the U.S.) mimic the effect of the hormone produced by the intestine. The hormone GLP-1 increases insulin production and release and decreases the level of glucagon, leading primarily to a reduction of postprandial glucose concentrations and HbA1c level (0.5-1.5 %) as well. In addition, they delay gastric emptying after the meals, what may contribute to their body weight-reducing effect observed during therapy (1.8–6.0 kg) [23, 26]. Pharmacologically, only short-acting GLP-1 analogs, like exenatide and lixisenatide, display a marked reduction of gastric emptying after the meals, while liraglutide and exenatide, formulated for slow release, have only a minor effect on it. Nevertheless, the mechanism of their action on GLP-1 receptors in the central nervous system which play a decisive role in energy homeostasis, has not yet been elucidated [27, 28]. Meta-analyses of 18 clinical trials report that GLP-1 receptor agonist groups for patients with diabetes achieved a greater weight loss than control groups (weighted mean difference -2.8 kg, 95 % confidence interval -3.4 to -2.3) [29].

SGLT2 INHIBITORS

Sodium-glucose linked co-transporter 2 (SGLT2) inhibitors (cana-, dapa-, empagliflozin) represent a new pharmacological group in the therapy of type 2 diabetes mellitus. The SGLT2 inhibitors exert their blood glucose-lowering and HbA1c reducing (0.4-1.0 %) effect by acting on the kidney, by increasing urinary glucose excretion, independently of insulin production and sensitivity (Fig 1). Results published up to now show that the body weight decreases during their use (0.7–1.48 kg) [8, 30-32]. The exact mechanism of this is yet unknown, however the results showed that SGLT2 therapy reduced the amount of total, visceral and subcutaneous adipose tissue as well [31].

ALPHA-GLUCOSIDASE INHIBITORS

By inhibiting the enzyme in the small intestine, alpha-glucosidase inhibitors (acarbose, miglitol) decrease the degradation to monosaccharides and consequently the absorption of carbohydrates taken in with the food. They cause a moderate reduction of postprandial blood glucose level and HbA1c concentration (0.5-0.8 %). Their use is neutral for body weight, or it is associated with its slight reduction [33].

SULFONYLUREA COMPOUNDS

Sulfonylureas (glyburide/glibenclamide/, glipizide, glimepiride and gliclazide /not licensed in the U.S.) stimulate insulin release from pancreatic β-cells *via* the adenosine triphosphate (ATP)-dependent potassium channels. They lower HbA1c level by 1.0 % to 2.0 %. The body weight may increase during their use (1.8–5 kg) [11, 23, 34]. As they act independently of actual blood glucose levels, they may increase the risk of developing hypoglycemia, and the consumption of plus carbohydrates for its correction. This may partly explain the increase of body weight. In addition it was suggested that they may also increase lipogenesis during their use [23].

GLINIDES

Similarly to the sulfonylureas, meglitinides (repa-, and nateglinide) enhance insulin secretion in the pancreatic β-cells, but their effect is rapid and short. They reduce primarily the postprandial glycemia and HbA1c level (0.5-1.5 %). The may cause a moderate increase of body weight (0.7–1.8 kg) due to the causes described under the sulfonylureas [23].

AMYLIN MIMETIC

The amylin receptor agonist pramlintide (not licensed in Europe for type 2 diabetes) decreases postprandial glucagon secretion, slows down gastric emptying, and increases satiety. It reduces primarily the postprandial glycemia and HbA1c level (0.5-0.7 %). Body weight may show a moderate decrease during its use [9].

DOPAMINE-2 AGONIST

Bromocriptine (quick release; not licensed in Europe for type 2 diabetes) activates dopaminergic receptors. Bromocriptine is thought to act on the circadian neuronal activities in the hypothalamus, to reset an abnormally elevated hypothalamic drive for increased plasma glucose, free fatty acids, and triglycerides in diabetic patients [9]. As for the body weight, its effect is neutral or it may cause a slight reduction of body mass.

BILE ACID SEQUESTRANT

The exact mode of action of colesevelam is not known. According to the available results, it may decrease hepatic glucose production and increase GLP-1 levels. It is neutral for the body weight.

INSULIN

These agents improve glycemic control and lower HbA1c by 1.5 % to 3.5 %. Administration of the prandial and basal insulin increases the risk of hypoglycemia and the rise of body weight.

During the administration of human basal insulin (neutral protamine Hagedorn /NPH/), the chances of hypoglycemia and growing body weight is higher in comparison to the analog basal insulins (glargine, detemir and degludec /not licensed in the U.S. /) [9, 35]. Of the two analog basal insulins detemir insulin leads to a lower increment of body weight according to certain results [36-38].

Degludec insulin is the newest basal analog. When injected to the tissues, it takes a multi-hexamer structure which makes a continuous, smooth insulin adsorption possible. Clinical studies have demonstrated that its use leads to a similar metabolic control than that of glargine insulin, at a decreasing risk of hypoglycemia [9, 39].

The body weight-increasing effect of insulin may be related to a reduction of glycosuria, to the way of subcutaneous administration which differs from the physiologic (intra-portal) one, as well as to the rise of the hypoglycemic events which are associated with an intake of plus carbohydrate during their correction. In addition, the different effects of analog and human basal insulins on the body weight may be explained with the varying effects exerted on lipogenesis, gluconeogenesis and on the satiety center in the central nervous system [23, 40].

CONCLUDING REMARKS

The prevalence of type 2 diabetes is increasing more and more worldwide. The majority of these patients are overweight or obese. Medicines with various points of attack have appeared in the therapy of type 2 diabetes in the recent years. In addition to the control of carbohydrate metabolism, the effect of the medicines on body weight is also an important aspect in the therapy of this disease.

CONFLICT OF INTEREST

The authors confirm that this chapter content has no conflict of interest.

ACKNOWLEDGEMENTS

Declared none.

REFERENCES

[1]   World Health Organization. Obesity and overweight. 2015. http://www.who.int/mediacentre/factsheets/fs311/en/
[2]   IDF Diabetes Atlas Group: Update of mortality attributable to diabetes for the IDF diabetes atlas: estimates for the year 2011. Diabetes Res Clin Pract 2013; 100: 277-9.
[3]   Whiting DR, Guariguata L, Weil C, Shaw J: IDF diabetes atlas: global estimates of the prevalence of diabetes for 2011 and 2030. Diabetes Res Clin Pract 2011; 94: 311-21.
[4]   World Health Organization. Diabetes. 2015. http://www.who.int/mediacentre/factsheets/fs312/en/
[5]   Field AE, Coakley EH, Must A. *et al*. Impact of overweight on the risk of developing common chronic diseases during a 10-year period. Arch Intern Med 2001; 161: 1581-1586.
[6]   Sharma A, Lavie CJ, Borer JS. *et al*. Meta-analysis of the relation of body mass index to all-cause and cardiovascular mortality and hospitalization in patients with chronic heart failure. Am J Cardiol 2015; 115: 1428-1434.

[7]   Calabia J, Arcos E, Carrero JJ, Comas J, Vallés M. Does the obesity survival paradox of dialysis patients differ with age? Blood Purif 2015; 39: 193-199

[8]   American Diabetes Association. Approaches to glycemic treatment. Sec. 7. In standards and medical care in diabetes – 2015. Diabetes Care 2015; 38 (Suppl. 1): S41-'48.

[9]   Inzucchi SE, Bergenstal RM, Buse JB. *et al.* Management of hyperglycemia in type 2 diabetes, 2015: a patient-centered approach: update to a position statement of the american diabetes association and the european association for the study of diabetes. Diabetes Care 2015; 38: 140-149.

[10]   Viollet B, Guigas B, Sanz Garcia N, Leclerc J, Foretz M, Andreelli F. Cellular and molecular mechanisms of metformin: an overview. Clin Sci 2012; 122: 253-270.

[11]   Nichols GA, Gomez-Caminero A. Weight changes following the initiation of new anti-hyperglycaemic therapies. Diabetes Obes Metab 2007; 9: 96-102.

[12]   Johansen K. Efficacy of metformin in the treatment of NIDDM: meta-analysis. Diabetes Care 1999; 22: 33-37.

[13]   Stumvoll M, Nurjhan N, Perriello G, Dailey G, Gerich JE. Metabolic effects of metformin in non-insulin-dependent diabetes mellitus. N Engl J Med 1995; 333: 550-554.

[14]   Malin SK, Kashyap SR. Effects of metformin on weight loss: potential mechanims. Curr Opin Endicrinol Diabetes Obes 2014; 21: 323-329.

[15]   Lv WS, Wen JP, Li L. *et al.* The effect of metformin on food intake and its potential role in hypothalamic regulation in obese diabetic rats. Brain Res 2012; 1444: 11-19.

[16]   O'Neill H. AMPK and exercise: glucose uptake and insulin sensitivity. Diabetes Metab J 2013; 37: 1-21.

[17]   Lindsay JR, Duffy NA, McKillop AM, Ardill J, O'Harte FP, Flatt PR. Inhibition of dipeptidyl peptidase IV activity by oral metformin in type 2 diabetes. Diabetic Med 2005; 22: 654-657.

[18]   Shin NR, Lee JC, Lee HY. *et al.* An increase in the Akkermansia spp. population induced by metformin treatment improves glucose homeostasis in diet-induced obese mice. Gut 2014; 63: 727-735.

[19]   Napolitano A, Miller S, Nicholls AW. *et al.* Novel gut-based pharmacology of metformin in patients with type 2 diabetes mellitus. PloS ONE 2014; 9: e100778.

[20]   Jenkins N, Padilla J, Arce-Esquivel A. *et al.* Effects of endurance exercise training, metformin, and their combination on adipose tissue leptin and IL-10 secretion in OLETF rats. J Appl Physiol 2012; 113: 1873-1883.

[21]   Woo SL, Xu H, Li H. *et al.* Metformin amilorates hepatic steatosis and inflammation without altering adipose phenotype in diet-induced obesity. PLoS One 2014; 9: e91111.

[22]   Kubota N, Yamauchi T, Tobe K, Kadowaki T. Adiponectin-dependent and –independent pathways in insulin-sensitizing and antidiabetic actions of thiazolidinediones. Diabetes 2006; 55 (Suppl2): S32-S38.

[23]   Siram AT, Yanagisawa R, Skamagas M. Weight management in type 2 diabetes mellitus. Mt Sinai J Med 2010; 77: 533-548.

[24]   Retnakaran R, Ye C, Hanley AJ, Harris SB, Zimman B. Discordant effects on central obesity, hepatic insulin resistance, and alanine aminotransferase of low-dose metformin and thiazolidinedione combination therapy in patients with impaired glucose tolerance. Diabetes Obes Metab 2012; 14: 91-3.

[25]   Vasudevan AR, Balasubramanyam A. Thiazolidinediones: a review of their mechanisms of insulin sensitization, therapeutic potential, clinical efficacy and tolerability. Diabetes Technol Ther 2004; 6: 850-863.

[26]   Pratley R, Nauck M, Bailey T. *et al.* One year of liraglutide treatment offers sustained and more effective glycaemic control and weight reduction compared with sitagliptin, both in combination

with metformin, in patients with type 2 diabetes: a randomised, parallel-group, open-label trial. Int J Clin Pract 2011; 65: 397-407.

[27]   Secher A, Jelsing J, Baquero AF. *et al.* The arcuate nucleus mediates GLP-1 receptor agonist liraglutid-dependent weight loss. J Clin Invest 2014; 124: 4473-4488.

[28]   Vrang N, Larsen PJ. Preproglucagon derived peptides GLP-1, GLP-2 and oxyntomodulin in the CNS: role of peripherally secreted and centrally produced peptides. Prog Neurobiol 2010; 92: 442-462.

[29]   Vilsbøll T, Christensen M, Junker AE, Knop FK, Gluud LL. Effects of glucagon-like peptide-1 receptor agonists on weight loss: systematic review and meta-analyses of randomised controlled trials. BMJ 2012; 344: d7771.

[30]   Rosenstock J, Vico M, Wei L, Salsali A, List JF. Effects of dapagliflozin, an SGLT2 inhibitor, on HbA1c, body weight, and hypoglycemia risk in patients with type 2 diabetes inadequately controlled on pioglitazone monotherapy. Diabetes Care 2012; 35: 1473-1478.

[31]   Bollinder J, Ljunggren Ö, Kullberg J. *et al.* Effects of dapagliflozin on body weight, total fat mass, and regional adipose tissue distribution in patients with type 2 diabetes mellitus with inadequate glycemic control on metformin. J Clin Endocrinol Metab 2012; 97: 1020-1031.

[32]   Goring S, Hawkins N, Wygant G. *et al.* Dapagliflozin compared with other anti-diabetes treatments when added to metformin monotherapy: a systematic review and network meta-analysis. Diabetes Obes Metab 2014; 16: 433-442.

[33]   Van de Laar FA, Lucassen PL, Akkermans RP, Van de Lisdonk EH, Rutten G, Van Weel C. Alpha-glucosidase inhibitors for patients with type 2 diabetes: results from a Cochrane systematic review and meta-analysis. Diabetes Care 2005; 28: 154-163.

[34]   UKPDS Group. United Kingdom Prospective Diabetes Study 24: a 6-year, randomized, controlled trial comparing sulfonylurea, insulin, and metformin therapy in patients with newly diagnosed type 2 diabetes that could not be controlled with diet therapy. Ann Intern Med 1998; 128: 165-175.

[35]   Montañana CF, Herrero CH, Fernández MR. Less weight gain and hypoglycaemia with once-daily insulin detemir than NPH insulin in intensification of insulin therapy in overweight type 2 diabetes patients – the PREDICTIVE BMI clinical trial. Diabet Med 2008; 25: 916-923.

[36]   Hollander P, Cooper J, Bregnhøj J, Pedersen CB. A 52-week, multinational, open-label, parallel-group, noninferiority, treat-to-target trial comparing insulin detemir with insulin glargine in a basal-bolus regimen with mealtime insulin aspart in patients with type 2 diabetes. Clin Ther 2008; 30: 1976-1987.

[37]   Rosenstock J, Davies M, Home PD, Larsen J, Koenen C, Schernthaner G. A randomised, 52-week, treat-to-target trial comparing insulin detemir with insulin glargine when administered as add-on to glucose-lowering drugs in insulin-naive people with type 2 diabetes. Diabetologia 2008; 51: 408-416.

[38]   Raslová K, Tamer SC, Clauson P, Karl D. Insulin detemir results in less weight gain than NPH insulin when used in basal-bolus therapy for type 2 diabetes mellitus, and this advantage increases with baseline body mass index. Clin Drug Investig 2007; 27: 279-285.

[39]   Rodbard HW, Cariou B, Zinman B. *et al.* Comparison of insulin degludec with insulin glargine in insulin-naive subjects with type 2 diabetes: a 2-year randomized, treat-to-target trial. Diabet Med 2013; 30: 1298-1304.

[40]   Hallschmid M, Jauch-Chara K, Korn O. *et al.* Euglycemic infusion of insulin detemir compared with human insulin appears to increase direct current brain potential response and reduces food intake while inducing similar systemic effects. Diabetes 2010; 59: 1001-1107.

CHAPTER 3

Sodium-Glucose Co-Transport Inhibitors: A New Therapeutic Option for Type 2 Diabetes Mellitus

Athanasia K. Papazafiropoulou[1,*] and Marina S. Kardara[2]

[1] 1st Department of Internal Medicine & Diabetes Center General Hospital of Piraeus "Tzaneio", Greece and [2] Health Center of Erymantheia, Achaia, Greece

Abstract: It is well established that kidneys play a major role to the homoeostasis of glucose at the level of production as well as at the level of absorption. Usually, the majority of the glucose amount that inserts the kidney's circulation is reabsorbed in the tubular system maintaining in this way a stable glucose level in the plasma. However, in some situations, when the kidneys are not able to reabsorb the whole amount of glucose, the excess glucose is excreted in urine. The last situation is called glycosuria. Sodium-coupled glucose transporters (SGLTs), especially SGLT2, have a key role to the glucose reabsorption in the tubular system and recently have been recognized as a target for the development of a new category of antidiabetic agents. A lot of clinical studies with agents that inhibit SGLT2 showed a favorable effect to plasma glucose levels as well as to body weight. Urinary tract infections are the more important adverse event of this category of antidiabetic agents without any significant implications.

Keywords: Kidneys, glucose reabsorption, phlorizin, dapagliflozin, sergliflozin, and type-2 sodium glucose co transporter.

INTRODUCTION

It is well established that type 2 diabetes mellitus (T2DM) is a worldwide epidemic taking increased dimensions [1]. Furthermore, T2DM is associated with increased morbidity and mortality due to diabetic complications, microvascular and macrovascular [1]. Hyperglycemia is the main pathogenetic risk factor for microvascular complications [2]. Therefore, achieving normal ranges of blood glucose levels is the key of the therapeutic management of the diabetic patient in order to prevent the diabetic complications, especially macrovascular disease [2].

*Address correspondence Athanasia K. Papazafiropoulou: 1st Department of Internal Medicine & Diabetes Center General Hospital of Piraeus "Tzaneio", Greece; Tel: 0030 697 9969483; Fax : 0030 2104592374; E-mail: pathan@ath.forthnet.gr

Atta-ur-Rahman (Ed)

Until now the existing antidiabetic agents include insulin and biguanides, thiazolidenediones, sulfonylureas, alpha-glucosidase inhibitors and the newest category, that use the incretin effect (glucagon-like peptide-1 analogues and dipeptidyl peptidase 4 inhibitors). These agents may be associated with significant side effects including hypoglycemia (sulfonylureas, insulin), weight gain (sulfonylureas, thiazolidenediones, and insulin), edema (thiazolidenediones), and possibly adverse cardiovascular outcomes (thiazolidenediones). More important is the fact that despite the use of the above medications the majority of the diabetic patients are outside the HbA1c targets [3, 4]. Therefore, still remains the need for novel antidiabetic agents, and inhibition of sodium-coupled glucose transporters (SGLTs) is a promising choice.

Role of Sodium Glucose Transporter in Renal Glucose Excretion

It is well established that kidneys play a major role to the homoeostasis of glucose at the level of production as well as at the level of absorption. Usually, the majority of the glucose amount (99%) that inserts the kidney's circulation (180 g/day) is reabsorbed in the tubular system maintaining in this way a stable glucose level in the plasma. However, in some situations, when the kidneys are not able to reabsorb the whole amount of glucose, the excess glucose is excreted in urine. The last situation is called glycosuria [5]. The sodium-coupled glucose transporters (SGLTs) are membrane proteins involved in the reabsorption of glucose [6]. SGLT1 and SGLT2 are large-membrane proteins (670 amino acids), and each has 14 transmembrane domains. The homology between SGLT1 and SGLT2 is approximately 58% [6, 7]. SGLT2s are located in the kidneys and they have the main role to glucose reabsorption [6, 7].

SGLT2 has low affinity and high capacity for glucose transport. SGLT1 has high affinity but low capacity for glucose transport [8]. Three additional SGLTs, SGLT3, SGLT4 and SGLT5, have been identified with a low-affinity transporter for glucose [6, 7]. SGLTs couple glucose reabsorption to sodium reabsorption, and in this way mediate renal glucose reabsorption. The sodium-potassium adenosine triphosphatase pump transports sodium across the basolateral surface into the intracellular fluid, maintaining the physiological levels of sodium in the cell. The inward sodium concentration gradient drives the glucose reabsorption.

The sodium electrochemical gradient generated by active sodium transport provides the energy required for glucose transport [5, 6]. The early convoluted segment (S1) of the proximal tubule reabsorbs approximately 90% of the filtered renal glucose. This is accomplished by the high- capacity, low-affinity SGLT2 transporter. The remaining 10% of the filtered glucose is reabsorbed by the high-affinity, low-capacity SGLT1 transporter in the distal straight segment (S3) of the proximal tubule [6-8]. During periods of hyperglycemia, the glucose reabsorptive capacity of the kidney increases in proportion to the plasma glucose concentration. However, as plasma glucose concentrations increase, the filtered glucose load increases in a linear manner. When the rate of glucose entering the nephron rises above 260-350 mg/min/1.73 m^2, for example in subjects with diabetes, the excess glucose outstrips resorptive capacity and appears in the urine [9]. In healthy subjects, this equates to a blood glucose concentration of approximately 200 mg/dl [10].

The idea for the use of SGLT2 inhibition for the treatment of diabetes came from a disease known as familial renal glucosuria [11]. Familial renal glucosuria is characterized by urinary glucose excretion in the presence of a normal blood glucose concentration and the absence of renal tubular dysfunction [11]. Subjects with familial renal glucosuria carry mutations in the genes that codes SGLT2 (SLC5A2) [11]. The majority of SGLT2 mutations are nonsense that result in disruption of transmembrane domains 10–13, which are essential for sugar transport by the SGLT2 [11]. The transmission of this rare disease is thought to be co-dominant with incomplete penetrance. Affected subjects with loss-of-function mutations in the gene encoding for the SGLT2 transporter manifest varying degrees of glucosuria (20–200 g/day). However, despite the glucosuria, these subjects are asymptomatic and have no known abnormalities of glucose or renal function [12]. They have not demonstrated an increased incidence of diabetes, chronic kidney disease, or urinary tract infection and have normal life expectancy [12]. On the other hand, subjects with the SGLT-1 mutation develop glucose-galactose malabsorbtion causing profuse diarrhea. These subjects have only mild glycosuria [11, 12].

Experimental studies in mice lacking the SGLT2 transporter showed glucosuria, polyuria, and increased food and fluid intake. However, SGLT2 knockout mice had comparable body weight, plasma glucose concentration, glomerular filtration rate (GFR), and urinary excretion of electrolytes and amino acids [13]. In humans, the same results were found; inhibition of SGLT2 in diabetics resulted in plasma glucose lowering [14, 15]. Except for plasma glucose lowering, SGLT2 inhibition has pleiotropic effects including a reduction in hepatic gluco-neogenesis [16] and a reduction of arterial blood pressure [17, 18].

Development of inhibitors of SGLTs Since the start of the 20th century, phlorizin, isolated in 1835 by a French chemist from the bark of apple trees, has been known to increase glycosuria [19, 20]. Phlorizin is comprised of a glucose ring connected *via* an oxygen atom to two phenol rings. Intravenous injection of phlorizin in normal subjects produces glucosuria, resembling familial renal glucosuria [20]. Later on, studies focused on the cellular mechanism of phlorizin action and showed that phlorizin inhibits glucose transport in the kidney and small intestine [20, 21]. It has been found that phlorizin inhibits both SGLT1 and SGLT2 in the proximal tubule but with a higher affinity for the SGLT2 transporter [20, 21]. In diabetic animal models, phlorizin increased glucose secretion in urine, caused a normalizing of plasma glucose, without inducing hypoglycemia [22]. Despite the efficacy of phlorizin in inhibiting SGLT2 transporter in diabetic animals, its clinical usefulness is negated by the low bioavailability following oral administration and the inhibition of SGLT1 in the gastrointestinal tract [22]. Furthermore, a metabolite of phlorizin, phloretin, inhibits both GLUT2- and GLUT1-mediated glucose absorption across the brush-border membrane of the gut [22]. Because of above limitations of phlorizin, other compounds with greater bioavailability after oral administration and higher selectivity for SGLT2 are, now, under development. The first two SGLT2 inhibitors, T-1095 and sergliflozin, have been discontinued after phase II clinical trials.

Dapagliflozin

Dapagliflozin can be administered once-daily and produces a significant glycosuria resulting in lowering plasma glucose levels [23]. It has 84% bioavailability in rats and a pharmacological half-life of 4.6 h [23]. Dapagliflozin is rapidly absorbed after oral administration, circulates bound to albumin and maximum plasma levels are showed 2 hours after oral administration [23].

Dapagliflozin as Monotherapy

In a 14-day study [24], conducted to evaluate glucosuria and glycemic parameters in T2DM subjects (5 mg, 25 mg, 100 mg of dapagliflozin *vs.* placebo) showed a significant reduction in fasting plasma glucose (FPG) were observed on day 2 with 100 mg dapagli-flozin (-9.3%, P < 0.001), and dose-dependent reductions were observed on day 13 with the 5 mg (-11.7%, P < 0.05), 25 mg (-13.3%, P < 0.05), and 100 mg (-21.8%, P < 0.0001) doses as compared with placebo. Significant improvements in oral glucose tolerance test were observed with all doses on days 2 and 13 (P < 0.001 as compared with placebo). On day 14, urine glucose values were 36.6-, 70.1-, and 69.9 g/day for the 5-, 25-, and 100 mg doses (as compared with no change for placebo), which were slightly lower than those on day 1. This was attributed to the decrease in filtered glucose load following improved glycemic control. The results of the study showed that dapagliflozin produced dose-dependent increases in glucosuria and clinically meaningful changes in glycemic parameters in T2DM patients [24].

Dapagliflozin at doses of 1-, 2.5- and 5 mg/d was effective in reducing glycemic levels and body weight in treatment-naïve patients with T2DM [25]. This phase III, randomized, double-blind, placebo-controlled study assigned 282 treatment-naïve patients with inadequate glycaemic control (HbA1c ≥7.0 and ≤10.0%) to placebo or dapagliflozin monotherapy (1-, 2.5- or 5 mg) daily for 24 weeks. At week 24, mean HbA1c reduction was significantly greater with dapagliflozin: -0.68% for 1 mg, -0.72% for 2.5 mg, -0.82% for 5 mg, *vs.* 0.02% for placebo (P < 0.0001). Mean FPG reduction was significantly greater for all dapagliflozin groups *vs.* placebo (P < 0.02), as was mean weight reduction (P < 0.003). Percentages of patients experiencing any ad- verse event were similar across groups [25].

In another study, compared with placebo, dapagliflozin (1-, 2.5-, 5- or 10 mg/d) significantly reduced hyperglycemia over 12 weeks with a low risk of hypoglycemia in 279 Japanese T2DM patients with inadequate glycemic control [29]. Significant reductions in HbA1c were seen with all dapagliflozin doses (-0.11 to -0.44%) *vs.* placebo (+0.37%). Reductions were also observed in FPG with dapagliflozin (-15.61 to -31.94 mg/dl) *vs.* placebo (+11.17 mg/dl). No significant difference in the proportion of patients achieving HbA1c levels <7.0% was noted with dapagliflozin *vs.* placebo. Adverse events were more frequent with dapagliflozin (40.7 to 53.8%) *vs.* placebo (38.9%) and mostly mild/moderate in intensity. The frequency of urinary tract or genital infections was 0 to 3.8% and 0 to 1.8% respectively with dapagliflozin and 1.9% and 0% with placebo [26].

In a 24-week parallel-group, double-blind, placebo-controlled phase III trial the efficacy and safety of dapagliflozin were evaluated in treatment-naïve patients with T2DM [27]. 485 patients with HbA1c 7.0-10.0% were randomly assigned to one of seven arms to receive once-daily placebo or 2.5-, 5-, or 10 mg dapagliflozin once daily in the morning or evening. Patients with HbA1c 10.1-12.0% were randomly as-signed to receive blinded treatment with a morning dose of 5- or 10 mg/d dapagliflozin. In the main cohort, mean HbA1c changes from baseline at week 24 were - 0.23% with placebo and -0.58, -0.77 (P = 0.0005 *vs.* placebo), and -0.89% (P < 0.0001 *vs.* placebo) with 2.5-, 5-, and 10 mg dapagliflozin, respectively. Signs, symptoms, and other reports suggestive of urinary tract infections and genital infection were more frequently noted in the dapagliflozin arms. There were no major episodes of hypoglycemia [27].

In a multiple-dose study safety and efficacy of dapagliflozin were evaluated in patients with T2DM [28]. Participants were randomly assigned to one of five dapagliflozin doses, metformin XR, or placebo for 12 weeks. After 12 weeks, dapagliflozin induced moderate glucosuria (200-300 kcal/day) and demonstrated significant glycemic improvements *vs.* placebo (HbA1c -0.55 to - 0.90% and FPG -16 to -31 mg/dl). Weight loss change *vs.* placebo was -1.3 to -2.0 kg. There was no change in renal function while serum uric acid decreased, serum magnesium increased, serum phosphate increased at higher doses, and dose-related 24-hours urine volume and hematocrit increased, all of small magnitude. Treatment-emergent adverse events were similar across all groups. Dapagliflozin improved hyperglycemia and induced weight loss in T2DM patients with no renal adverse events [28].

Dapagliflozin as Add-on Therapy

Two randomized, 24-week trials in treatment-naïve patients were conducted to compare dapagliflozin plus metformin, dapagliflozin alone and metformin alone in treatment-naïve patients with baseline HbA1c 7.5-12.0% [29]. Each trial had three arms: dapagliflozin plus metformin, dapagliflozin monotherapy and metformin monotherapy. Dapagliflozin in combination and as monotherapy was dosed at 5 mg (Study 1) and 10 mg (Study 2). Metformin in combination and as monotherapy was titrated to 2,000 mg. In both trials, combination therapy led to significantly greater reductions in HbA1c compared with either monotherapy: -2.05% for dapagliflozin plus metformin, -1.19% for dapagliflozin, and -1.35% for metformin (P < 0.0001) (Study 1); -1.98% for dapagliflozin plus metformin, -1.45% for dapagliflozin and -1.44% for metformin (P < 0.0001) (Study 2). Combination therapy was statistically superior to monotherapy in reduction of FPG (P < 0.0001 for both studies) and more effective than metformin for weight reduction (P < 0.0001). Dapagliflozin 10 mg was non-inferior to metformin in reducing HbA1c (Study 2). Events suggestive of genital infection were reported in 6.7%, 6.9% and 2.0% (Study 1) and 8.5%, 12.8% and 2.4% (Study 2) of patients in combination, dapagliflozin and metformin groups; events suggestive of urinary tract infection were reported in 7.7%, 7.9% and 7.5% (Study 1) and 7.6%, 11.0% and 4.3% (Study 2) of patients in the respective groups. No major hypoglycemia was reported. The results of the studies showed that in treatment-naïve patients with T2DM, dapagliflozin plus metformin was generally well tolerated and effective in reducing HbA1c, FPG and weight. Dapagliflozin-induced glucosuria led to an increase in events suggestive of urinary tract and genital infections [29].

In a phase III, placebo-controlled trial, 546 adults with T2DM who were receiving daily metformin (>/=1500 mg/day) and had inadequate glycemic control were randomly assigned to receive one of three doses of dapagliflozin (2.5-, 5-, or 10 mg) or placebo orally once daily [30]. At week 24, mean HbA1c had de-creased by -0.30% in the placebo group, compared with -0.67% (P =0.0002) in the dapagliflozin 2.5 mg group, -0.70% (P <0.0001) in the dapagliflozin 5 mg group, and - 0.84% (P <0.0001) in the dapagliflozin 10 mg group. Symptoms of hypoglycemia occurred in similar proportions of patients in the dapagliflozin (2-4%) and placebo groups (3%). Signs, symptoms, and other re-ports suggestive of genital infections were more frequent in the dapagliflozin groups than in the placebo group. 17 patients had serious adverse events (4 in each of the dapagliflozin groups and 5 in the placebo group) [30].

In a 52-week, noninferiority trial patients with T2DM (baseline mean HbA1c, 7.7%), who were receiving metformin monotherapy, were randomized to add-on dapagliflozin (n = 406) or glipizide (n = 408) up-titrated over 18 weeks, based on glycemic response and tolerability, to ≤10 or ≤20 mg/day, respectively [31]. The primary end point, adjusted mean HbA1c reduction with dapagliflozin (-0.52%) compared with glipizide (-0.52%), was statistically noninferior at 52 weeks. Key secondary end points: dapagliflozin produced significant adjusted mean weight loss (-3.2 kg) *vs.* weight gain (1.2 kg; P < 0.0001) with glipizide, significantly increased the proportion of patients achieving ≥5% body weight reduction (33.3%) *vs.* glipizide (2.5%; P < 0.0001), and significantly decreased the proportion experiencing hypoglycemia (3.5%) *vs.* glipizide (40.8%; P < 0.0001). Events suggestive of genital infections and lower urinary tract infections were reported more frequently with dapagliflozin compared with glipizide but responded to standard treatment and rarely led to study discontinuation [34]. Despite similar 52-week glycemic efficacy, dapagliflozin reduced weight and produced less hypoglycemia than glipizide in T2DM inadequately controlled with metformin [31].

A 24-week, multicentre trial enrolled patients with uncontrolled T2DM (HbA1c 7-10%) receiving sulphonylurea monotherapy [32]. Patients (n = 597) were randomly assigned to placebo or dapagliflozin (2.5-, 5- or 10 mg/day) added to open-label glimepiride 4 mg/day for 24 weeks. At 24-week, HbA1c adjusted mean changes from baseline for placebo *vs.* dapagliflozin 2.5/5/10 mg groups were -0.13 *vs.* -0.58, -0.63, -0.82%, respectively (all P < 0.0001 *vs.* placebo). Corresponding body weight and FPG values were -0.72, -1.18, -1.56, -2.26 kg and -0.11, -0.93, - 1.18, -1.58 mmol/l, respectively. In placebo *vs.* dapagliflozin groups, serious adverse events were 4.8 *vs.* 6.0-7.1%; hypoglycemic events 4.8 *vs.* 7.1-7.9%; events suggestive of genital infection 0.7 *vs.* 3.9-6.6%; and events suggestive of urinary tract infection 6.2 *vs.* 3.9-6.9%. No kidney infections were reported [35]. Dapagliflozin added in patients with T2DM uncontrolled on sulfonylurea monotherapy significantly improved HbA1c, reduced weight and was generally well tolerated, although events suggestive of genital infections were reported more often in the dapagliflozin group [32].

In patients with T2DM inadequately controlled on pioglitazone, the addition of dapagliflozin further reduced HbA1c levels and mitigated the pioglitazone-related weight gain without increasing hypoglycemia risk [33]. Treatment-naïve patients or those receiving metformin, sulfonylurea, or thiazolidinedione were randomized to 48 weeks of double-blind dapagliflozin 5 (n = 141) or 10 mg (n = 140) or placebo (n = 139) every day plus open-label pioglitazone. At week 24, the mean reduction from baseline in HbA1c was -0.42% for placebo *vs.* -0.82 and -0.97% for dapagliflozin 5- and 10 mg groups, respectively (P = 0.0007 and P < 0.0001 *vs.* placebo). Patients receiving pioglitazone alone had greater weight gain (3 kg) than those receiving dapagliflozin plus pioglitazone (0.7-1.4 kg) at week 48. Through 48 weeks: hypoglycemia was rare; more events suggestive of genital infection were reported with dapagliflozin (8.6-9.2%) than placebo (2.9%); events suggestive of urinary tract infection showed no clear drug effect (5.0-8.5% for dapagliflozin and 7.9% for placebo); dapagliflozin plus pioglitazone groups had less edema (2.1-4.3%) compared with placebo plus pioglitazone (6.5%); and congestive heart failure and fractures were rare [33].

A 24-week study evaluated the efficacy and safety of adding dapagliflozin therapy in T2DM patients inadequately controlled with insulin with or without oral antidiabetic drugs [34]. 808 T2DM patients receiving at least 30 units of insulin daily, with or without up to 2 oral antidiabetic drugs were randomly assigned to receive placebo or 2.5-, 5-, or 10 mg of dapagliflozin, once daily, for 48 weeks. After 24 weeks, mean HbA1c decreased by 0.79% to 0.96% with dapagliflozin compared with 0.39% with placebo in the 2.5 mg group, -0.49% in the 5 mg group, and -0.57% in the 10 mg group. Daily insulin dose decreased by 0.63 to 1.95 units with dapagliflozin and increased by 5.65 units with placebo. Body weight decreased by 0.92 to 1.61 kg with dapagliflozin and increased by 0.43 kg with placebo in the 2.5 mg group, -1.42 kg in the 5 mg group, and -2.04 kg in the 10 mg group. These effects were maintained at 48 weeks. Compared with the placebo group, patients in dapagliflozin groups had a higher rate of hypoglycemic episodes (56.6% *vs.* 51.8%), events suggesting genital infection (9.0% *vs.* 2.5%), and events suggesting urinary tract infection (9.7% *vs.* 5.1%) [37].

Dapagliflozin improved glycemic control, stabilized insulin dosing, and reduced weight without increasing major hypoglycemic episodes in insulin treated T2DM patients inadequately controlled [34]. In a multicenter study patients with T2DM that were poorly controlled with high insulin doses plus oral anti-diabetic agents (OADs) in the treatment cohort (n = 71) were randomly assigned to receive dapagliflozin at the dose of 10 mg or 20 mg. After 12 weeks, patients that took dapagliflozin at the dose of 10 mg or 20 mg demonstrated -0.70 and -0.78% mean differences in HbA1c change from baseline *vs.* placebo. In both dapagliflozin groups, 65.2% of patients achieved a decrease from baseline in HbA1c > or =0.5% *vs.* 15.8% in the placebo group. Mean changes from baseline in FPG were +17.8, +2.4, and -9.6 mg/dl (placebo, 10-, and 20 mg dapagliflozin, respectively). Post-prandial glucose reductions with dapagliflozin also showed dose dependence. Mean changes in total body weight were -1.9 kg (placebo), -4.5 kg (10 mg dapagliflozin), and -4.3 kg (20 mg dapagliflozin).

Overall, adverse events were balanced across all groups, although more infections of the genital system were observed in patients that took 20 mg of dapagliflozin. In patients receiving high insulin doses plus insulin sensitizers who had their baseline insulin reduced by 50%, dapagliflozin decreased HbA1c, produced better FPG and postprandial glucose levels, and lowered weight more than placebo [35]. In three-treatment crossover studies, 24 subjects received 50 mg dapagliflozin, 45 mg pioglitazone or the combination, while 18 subjects received 20 mg dapagliflozin, 1000 mg metformin or the combination. In an open-label, randomized, five-period, five-treatment, unbalanced crossover study, 18 subjects first received 20 mg dapagliflozin, 4 mg glimepiride or the combination, and afterward 100 mg sitagliptin or sitagliptin plus 20 mg dapagliflozin. Blood samples were taken over 72h of each treatment period. Co-administration of dapagliflozin with pioglitazone, metformin, glimepiride or sitagliptin had no effect on dapagliflozin maximum plasma concentration (Cmax) or area under the plasma concentration-time curve (AUC). Similarly, dapagliflozin did not affect the Cmax or AUC for the co-administered drug. All monotherapies and combination therapies were well tolerated. These studies showed that dapagliflozin can be co- administered with pioglitazone, metformin, glimepiride or sitagliptin without dose adjustment of either drug [36].

Canagliflozin

Canagliflozin is another selective SGLT2 inhibitor. In T2DM subjects, canagliflozin produces dose dependent glucosuria with a maximal effect at 400 mg/day [37]. Canagliflozin can lower plasma glucose levels as monotherapy and as add-on therapy to metformin and insulin [37].

In a 12-week study, canagliflozin monotherapy (300 mg/day) decreased HbA1c by 0.99% [38]. In another 12 weeks study in 451 T2DM subjects on metformin, canagliflozin (50, 100, 200, and 300 mg/day) reduced HbA1c by 0.7% to 0.9% from baseline in association with weight loss of 1.3 to 2.3 kg [39]. In 29 diabetic subjects treated with insulin, the addition of canagliflozin (100 and 300 mg/day) caused a 0.54% and 0.73% decrease in HbA1c after 28 days of treatment [40]. In a 16- day study canagliflozin was shown to improve β-cell function in T2DM subjects [41].

The efficacy and safety of canagliflozin monotherapy were evaluated over 52 weeks in patients with T2DM inadequately controlled with diet and exercise. This phase III study included a placebo-controlled, 26-week core period (canagliflozin 100 or 300 mg *vs.* placebo) and an active-controlled, 26-week extension (blinded switch of placebo-treated patients to sitagliptin 100 mg (placebo/sitagliptin)) [42].

At week 52, canagliflozin 100 and 300 mg provided dose-related decreases from baseline in HbA1c of -0.81% and -1.11%. Canagliflozin 100 and 300 mg decreased FPG (-1.5 and -2.2 mmol/L (-27.4 and -39.1 mg/dL)), body weight (-3.3% and -4.4%), and systolic blood pressure (-1.4 and -3.9 mmHg). Over 52 weeks, overall adverse events rates were 67.2%, 66.0%, and 64.1% with canagliflozin 100 and 300 mg and placebo/sitagliptin. Compared with placebo/sitagliptin, canagliflozin was associated with higher rates of genital mycotic infections and adverse events related to osmotic diuresis; these led to few discontinuations. Rates of volume depletion adverse events and documented hypoglycemia were low across groups [42]. Another phase III study in 1284 participants with T2DM aged 18 to 80 years who had inadequate glycaemic control (HbA1c 7.0 to 10.5%) on metformin therapy received canagliflozin 100 mg or 300 mg, sitagliptin 100 mg, or placebo for 52 weeks [43]. At week 26, canagliflozin 100 mg and 300 mg reduced HbA1c *vs.* placebo (-0.79%, -0.94%, -0.17%, respectively; p < 0.001). At week 52, canagliflozin 100 mg and 300 mg demonstrated non-inferiority, and canagliflozin 300 mg demonstrated statistical superiority, to sitagliptin in lowering HbA1c (-0.73%, -0.88%,-0.73%, respectively). Canagliflozin 100 mg and 300 mg reduced body weight *vs.* placebo (week 26: - 3.7%, -4.2%, -1.2%, respectively; p < 0.001) and sitagliptin (week 52: -3.8%, -4.2%, -1.3%, respectively; p < 0.001). Both canagliflozin doses reduced FPG and systolic blood pressure *vs.* placebo (week 26) and sitagliptin (week 52) (p < 0.001). Overall adverse events and adverse events -related discontinuation rates were generally similar across groups, but higher with canagliflozin 100 mg. Genital mycotic infection and osmotic diuresis-related adverse events rates were higher with canagliflozin; few led to discontinuations. Hypoglycaemia incidence was higher with canagliflozin [43].

In a 52 week, phase III non-inferiority trial 1452 patients aged 18-80 years with T2DM and HbA1c of 7.0-9.5% on stable metformin were randomly assigned (1:1:1) to receive canagliflozin 100 mg or 300 mg, or glimepiride (up-titrated to 6 mg or 8 mg per day) orally once daily [44]. For lowering of HbA1c at 52 weeks, canagliflozin 100 mg was non-inferior to glimepiride (-0.01%), and canagliflozin 300 mg was superior to glimepiride (-0.12%). Thirty nine (8%) patients had serious adverse events in the glimepiride group *vs.* 24 (5%) in the canagliflozin 100 mg group and 26 (5%) in the 300 mg group. In the canagliflozin 100 mg and 300 mg groups *vs.* the glimepiride group, a greater number of genital mycotic infections, urinary tract infections (UTIs), and osmotic diuresis-related events were recorded [44].

In a 52-week, phase III study, 755 diabetic subjects using stable metformin plus sulfonylurea received canagliflozin 300 mg or sitagliptin 100 mg daily. At 52 weeks, canagliflozin 300 mg demonstrated noninferiority and, in a subsequent assessment, showed superiority to sitagliptin 100 mg in reducing HbA1c (-1.03% and -0.66%, respectively) [45].

Greater reductions in FPG, body weight, and systolic blood pressure were observed with canagliflozin *vs.* sitagliptin (P < 0.001). Overall adverse events rates were similar with canagliflozin (76.7%) and sitagliptin (77.5%); incidence of serious adverse events and adverse events -related discontinuations was low for both groups. Higher incidences of genital mycotic infections and osmotic diuresis-related adverse events were observed with canagliflozin, which led to one discontinuation. Hypoglycemia rates were similar in both groups [45].

Another study was designed to evaluate the effects of canagliflozin in T2DM subjects inadequately controlled with metformin monotherapy. 451 subjects were randomized to canagliflozin 50, 100, 200, or 300 mg once daily (QD) or 300 mg twice daily (BID), sitagliptin 100 mg QD, or placebo [46]. Canagliflozin was associated with significant reductions in HbA1c from baseline (7.6-8.0%) to week 12: -0.79, -0.76, -0.70, -0.92, and -0.95% for canagliflozin 50, 100, 200, 300 mg QD and 300 mg BID, respectively, *vs.* -0.22% for placebo (all P < 0.001) and -0.74% for sitagliptin. FPG was reduced by -16 to -27 mg/dL, and body weight was reduced by - 2.3 to -3.4%, with significant increases in urinary glucose-to-creatinine ratio. Adverse events were transient, mild to moderate, and balanced across arms except for a non-dose-dependent increase in symptomatic genital infections with canagliflozin (3-8%) *vs.* placebo and sitagliptin (2%). Urinary tract infections were reported without dose dependency in 3-9% of canagliflozin, 6% of placebo, and 2% of sitagliptin arms. Overall incidence of hypoglycemia was low [46].

A 28-day study conducted in 29 subjects with T2DM not optimally controlled on insulin and up to one oral antihyperglycaemic agent were treated with canagliflozin 100 mg QD or 300 mg BID or placebo [47]. Canagliflozin pharmacokinetics was dose-dependent, and the elimination half-life ranged from 12 to 15 h. After 28 days, the renal threshold for glucose excretion was reduced; urinary glucose excretion was increased; and HbA1c, FPG and body weight decreased in subjects administered canagliflozin (HbA1c reductions: 0.19% with placebo, 0.73% with 100 mg QD, 0.92% with 300 mg BID; body weight changes: 0.03 kg increase with placebo, 0.73 kg reduction with 100 mg QD, 1.19 kg reduction with 300 mg BID). There were no serious adverse events or severe hypoglycaemic episodes. The incidence of adverse events was similar across groups. The study showed that in subjects receiving insulin and oral antihyperglycaemic therapy, canagliflozin was well tolerated showing had pharmacokinetic characteristics consistent with once-daily dosing, and improved glycaemic control [47].

Another phase III study in 716 subjects aged 55 to 80 years with HbA1c levels 7.0% to 10.0% were randomized to receive canagliflozin 100 mg or 300 mg or placebo (1:1:1) daily [48]. At week 26, treatment with canagliflozin 100 mg and

300 mg significantly reduced HbA1c levels compared with placebo (-0.60%, -0.73%, -0.03%, respectively; P < 0.001); more subjects achieved HbA1c levels < 7.0% with both canagliflozin doses compared with placebo (P < 0.001). Both canagliflozin doses significantly reduced body weight, FPG level, and systolic blood pressure, and increased high density lipoprotein cholesterol level compared with placebo (P < 0.001); low-density lipoprotein cholesterol level was increased with both canagliflozin doses compared with placebo. The overall adverse events incidence was slightly higher with canagliflozin 300 mg than with canagliflozin 100 mg or placebo (78.0%, 72.2%, and 73.4%, respectively). Serious adverse events and adverse events -related discontinuation rates were low across groups. Both canagliflozin doses were associated with higher rates than placebo of genital mycotic infections, UTIs, and osmotic diuresis-related adverse events. Documented hypoglycemia rates were modestly higher with both canagliflozin doses compared with placebo [48].

Another study evaluated the efficacy and safety of canagliflozin in subjects with T2DM and stage 3 chronic kidney disease (CKD; GFR ≥30 and <50 ml/min/1.73m^2) [49]. In this phase III trial, 269 subjects received canagliflozin 100 or 300 mg or placebo daily. Both canagliflozin 100 and 300 mg reduced HbA1c from baseline compared with placebo at week 26 (-0.33, -0.44 and - 0.03%; p < 0.05). Numerical reductions in FPG and higher proportions of subjects reaching HbA1c < 7.0% were observed with canagliflozin 100 and 300 mg *vs.* placebo (27.3, 32.6 and 17.2%). Overall adverse events rates were similar for canagliflozin 100 and 300 mg and placebo (78.9, 74.2 and 74.4%). Slightly higher rates of UTIs and adverse events related to osmotic diuresis and reduced intravascular volume were observed with canagliflozin 300 mg compared with other groups. Transient changes in renal function parameters that trended towards baseline over 26 weeks were observed with canagliflozin. The results of the study showed that canagliflozin improved glycaemic control and was generally well tolerated in subjects with T2DM and Stage 3 CKD [49].

In a phase II study, T2DM subjects with inadequate glycemic control on metformin were enrolled and randomized to one of seven arms - placebo; canagliflozin doses 50 -, 100 -, 200 -, 300 mg daily, or 300 mg twice daily; and sitagliptin 100 mg daily - for 12 weeks [50]. At the end of the study canagliflozin increased renal glucose excretion by 35.4-61.6 mg/mg creatinine in the five dose groups. In the placebo group renal glucose excretion was increased by 1.9 mg/mg creatinine, and in the sitagliptin group it decreased by 1.9 mg/mg creatinine. Asymptomatic bacteriuria were present in 6.4% of canagliflozin and 6.5% of placebo/sitagliptin (control) subjects at randomization and, at 12 weeks, in 7.7% and 6.3% of subjects, respectively (OR: 1.23; 95% CI: 0.45-3.89). For subjects with initially negative urine cultures at baseline, 3 out of 82 (3.7%) who received controls and 10 out of 207 (4.8%) who received canagliflozin developed bacteriuria (p = 0.76) at week 12. There were 21 adverse event reports of urinary tract infections; 16 (5.0%) in canagliflozin subjects and 5 (3.8%) in control subjects (OR: 1.31; 95% CI: 0.45-4.68). In this trial, when compared with control subjects, canagliflozin increased urinary glucose excretion but was not associated with increased bacteriuria or adverse events reports of UTIs [50].

In a double-blind study, subjects with T2DM and inadequate glycemic control on metformin were randomized to placebo; canagliflozin 50, 100, 200, 300 mg daily or 300 mg twice daily; or sitagliptin 100 mg daily for 12 weeks [51]. Vaginal swabs for Candida culture were collected from 198 female subjects at baseline and week 12, and during the trial if symptoms consistent with vulvovaginal candidiasis occurred. At baseline, 23/198 (12%) females had vaginal cultures positive for Candida, with age ≤55 years associated with increased risk (OR: 3.5; 95% CI: 1.1- 10.7). Of those with negative cultures at baseline, 31% of canagliflozin and 14% of placebo/sitagliptin subjects converted to positive at week 12 (OR: 2.8; 95% CI: 1.0-7.3 for canagliflozin *vs.* placebo/sitagliptin). Two placebo/sitagliptin (3%) and 16 canagliflozin subjects (10%) experienced vulvovaginal adverse events. Positive vaginal culture for Candida species at baseline was a risk factor for vulvovaginal adverse events (OR: 9.1; 95% CI: 2.4-34.0). All subjects in the canagliflozin group with a vaginal culture taken at the time of the vulvovaginal adverse events were positive for Candida species. Most vulvovaginal adverse events were treated with antifungal therapy and resolved without study drug interruption; none led to discontinuation. Study limitations include small population, short duration, and not obtaining cultures in all women with vulvovaginal adverse events [51].

Empagliflozin

Empagliflozin produces a dose-dependent glucosuria in T2DM subjects [52]. In a 4-week study, empagliflozin (100 mg/day) caused urinary glucosuria of 74 g/day [53]. In a placebo controlled 12- week study in 495 diabetic subjects with poor glycemic control on metformin, empagliflozin (25 mg/day) caused a decrease in fasting blood glucose levels and HbA1c [54]. Ipragliflozin In a study, 361 Japanese T2DM subjects treated with ipragliflozin at doses ranging from 12.5 to 100 mg/day experienced a 0.9% reduction in HbA1c followed by a reduction to body weight [55]. In another study in 62 T2DM subjects, ipragliflozin monotherapy (50 mg/day) caused a 1.2% decrease in HbA1c [56].

Remogliflozin Etabonate

Remogliflozin etabonate is another SGLT2 inhibitor that is under development. However, the findings of experimental studies are conflicting showing different results; some studies confirmed that remogliflozin etabonate causes significant glycosuria while there is one study in 6-hours fasted rats that showed no result in plasma glucose level [57]. In humans, a study in 13 patients with T2Dm showed that remogliflozin etabonate, as add-on therapy to metformin, had a favorable effect to glucose levels and neutral effect to the onset of hypoglycemic episodes [58]. The favorable effect of remogliflozin etabonate to glucose levels was confirmed by another study in 6 subjects with T2DM where [59].

Safety

According to the data of the existing clinical studies with the use of SGLT2 inhibitors the most observed adverse events were urinary tract infections (UTIs) and vulvovaginal infections [60]. In clinical studies, a small (3–5%) increase in the rate of UTIs has been reported in subjects receiving SGLT2 inhibitors compared to placebo [60]. The majority of these infections involved cystitis, vulva- vaginitis and balanitis and have responded to standard antibiotic and local antifungal treatment [60]. In trials where dapagliflozin was used there was a small increased to the incidence of UTIs compared with placebo and metformin [17]. However, studies have showed that diabetics, especially diabetic women, develop more frequent than the general population UTIs and recurrent vaginal candidiasis

[60]. However, a prospective study showed that in diabetic women glycosuria is not associated with increased risk of UTIs [61].

Another risk with treatment with SGLT2 inhibitors is a possible increase in urine volume as well as loss of electrolytes, and a 400 to 500 ml negative fluid balance occurs during the first 2–3 days of therapy. However, when dapagliflozin was administered to humans, urine volume increased only modestly during the first 2 to 3 days after initiation of therapy. Excessive urine loss of sodium, potassium, and other electrolytes was not observed [17]. This mild volume contraction was followed by a small rise in hematocrit and plasma urea nitrogen to creatinine ratio as well as a decrease in blood pressure. Finally, plasma electrolyte concentrations did not change in dapagliflozin group [17, 60]. Tachycardia and orthostatic hypotension (clinical signs of volume depletion) and hyponatremia (laboratory evidence of water depletion) have not been reported in subjects treated with SGLT2 inhibitors [17, 60]. According to studies, SGLT2 inhibitors do not have any deleterious effect on renal function in subjects with T2DM with normal levels of GFR, electrolyte disturbances, acid-base balance, hypertension and patient's quality of life [62]. At this point it must be mentioned that all these studies were performed in subjects with normal renal function, and further studies are needed in order to clarify the effects of SGLT2 inhibitors in subjects with impaired renal function.

An increased incidence of bladder and breast cancer was observed in phase III studies of dapagliflozin [62]. There were 9 cases of breast cancer in the dapagliflozin group (2223 patients) compared to 1 case in placebo (1053 patients), all diagnosed within the first year of the studies [62]. Bladder cancers were reported in 9 cases in the dapagliflozin group (5478 patients) compared to 1 case in the control group (3156 patients) [62]. All were men and 6 patients had a history of hematuria before receiving the study drug [62]. In preclinical studies of dapagliflozin in rodents there was no evidence of carcinogenicity [63]. It is important to notice that the significance of the increased incidence of these tumors observed in dapagliflozin studies remains uncertain and further studies are needed to be determined [62].

Other Effects

Preclinical and early clinical studies showed that SGLT2 inhibitors do not cause hypoglycaemia [5]. This is explained by the fact that SGLT2 inhibitors decrease the plasma glucose concentration without augmenting insulin secretion by the pancreatic β-cells and because of the renal threshold of glycaemia, below which SGLT2 inhibitors would not be expected to cause further urinary glucose excretion [5]. According to the data by the clinical trials, the prevalence of hypoglycemic events in subjects treated with SGLT2 inhibitors was similar to that in people receiving placebo [5]. However, when SGLT2 inhibitors are used in combination with sulfonylurea or insulin, physicians should consider reducing the dose of sulfonylurea or insulin in order to avoid possible hypoglycemic events.

Another observation in SGLT2 inhibitors studies is the weight loss. It is known that urinary loss of 60 to 80 g of glucose per day equates to 240 to 320 cal/day, resulting in the weight loss that is observed in all clinical studies with SGLT2 inhibitors [17, 60]. As it is previously mentioned early decreases in weight may be the result of the osmotic diuretic effect of the agents, whereas weight loss over subsequent weeks may be the result of caloric loss [17, 60]. Weight loss of 2–3 Kg has been demonstrated in 12 week trials of dapagliflozin, canagliflozin and empagliflozin [17, 60]. Another finding in dapagliflozin studies is the mild reduction in systolic and diastolic blood pressure [17] that is attributed to the fluid/sodium deficit that occurs during the first several days of dapagliflozin treatment [17, 60].

Small reductions in GFR occur shortly after the initiation of the therapy, returning to normal after a few weeks [62, 64]. In addition, because of their mechanism of action, the efficacy of SGLT2 inhibitors to reduce the plasma glucose levels is highly dependent upon renal function. In subjects with GFR levels between 60–90 ml/min, the glucosuria produced by dapagliflozin [62] was decreased by 40% and the reduction in HbA1c was decreased by about 20%. Among subjects with similarly impaired renal function, ipragliflozin was reported to produce comparable glucosuria to subjects with GFR greater than 90 ml/min [64]; however, the decrease in fasting blood glucose levels was decreased by 50%. In subjects with GFR levels between 30–59 ml/min, the glucosuria produced by both ipragliflozin and dapagliflozin was reduced but the decrease in fasting blood glucose levels and HbA1c was clinically insignificant.

CONCLUSION

Current data in experimental models and humans show that SGLT2 inhibitors are an effective treatment to reduce plasma glucose levels in T2DM subjects with a good safety profile. SGLT2 inhibitors can be used in combination with all other antidiabetic medications. Because the SGLT2 inhibitors have a mechanism of action that is independent of insulin secretion or insulin resistance, their efficacy is not declined with progressive β-cell failure, although it will decline with declining renal function. The mode of action of SGLT2 inhibitors potentially makes them an option at any stage in the disease process and in any combination. Given all these data inhibition of the SGLT2 transporter is a promising treatment for T2DM subjects and has to be proved in clinical practice.

ACKNOWLEDGEMENTS

Declared None.

CONFLICT OF INTEREST

The authors have declared that no competing interests exist.

REFERENCES

[1]   International Diabetes Federation (IDF). Diabetes Atlas. Available at: http://www.diabetesatlas.org. (Accessed on: 2012).
[2]   American Diabetes Association. Diagnosis and classification of diabetes mellitus. Diabetes Care 2006; 29(Suppl 1): S43-S48.
[3]   Fan T, Koro CE, Fedder DO, Bowlin SJ. Ethnic disparities and trends in glycemic control among adults with type 2 diabetes in the U.S. from 1988 to 2002. Diabetes Care 2006; 29: 1924-5.
[4]   Koro CE, Bowlin SJ, Bourgeois N, Fedder DO. Glycemic control from 1988 to 2000 among US adults diagnosed with type 2 diabetes: a preliminary report. Diabetes Care 2004; 27:17-20.
[5]   Isaji M. Sodium-glucose cotransporter inhibitors for diabetes. Curr Opin Investig Drugs 2007; 8: 285-92.
[6]   Wright EM, Turk E. The sodium/glucose cotransport family SLC5. Pflugers Arch 2004; 447: 510-18.

[7]   Kanai Y, Lee WS, You G, Brown D, Hediger MA. The human kidney low affinity Na/glucose cotransporter SGLT2. Delineation of the major renal reabsorptive mechanism for D-glucose. J Clin Invest 1994; 93: 397-404.

[8]   Valtin H. Tubular reabsorption. In renal function. Boston: Little, Brown and Company; 1983.

[9]   Zelikovic I. Aminoaciduria and glycosuria. In: Avner ED, Harmon WE and Niaudet P, eds. Pediatric Nephrology. 5th edition. Philadelphia: Lippincott Williams & Wilkins: 2004; 701-28.

[10]   Moe OW, Wright SH and Palacín M. Renal handling of organic solutes. In: Brenner BM, ed. Brenner and Rector's The Kidney. 8th edition. Philadelphia: Saunders Elsevier: 2008; 214-7.

[11]   Santer R, Calado J. Familial renal glucosuria and SGLT2: from a Mendelian trait to a therapeutic target. Clin J Am Soc Nephrol 2010; 5: 133-41.

[12]   Santer R, Kinner M, Lassen CL, *et al.,* Molecular analysis of the SGLT2 gene in patients with renal glucosuria. J Am Soc Nephrol 2003; 14: 2873-82.

[13]   Vallon V, Platt KA, Cunard R, *et al.,* SGLT2 mediates glucose reabsorption in the early proximal tubule. J Am Soc Nephrol 2011; 22: 104-12.

[14]   Han S, Hagan DL, Taylor JR, *et al.,* Dapagliflozin, a selective SGLT-2 inhibitor, improves glucose homeostasis in normal and diabetic rats. Diabetes 2008; 57: 1723-9.

[15]   Jabbour SA, Goldstein BJ. Sodium-glucose cotransporter 2 inhibitors: blocking renal tubular reabsorption of glucose to improve glycaemic control in patients with diabetes. Int J Clin Pract 2008; 62: 1279-84.

[16]   Kahn BB, Shulman GI, DeFronzo RA, Cushman SW, Rossetti L. Normalization of blood glucose in diabetic rats with phlorizin treatment reverses insulin-resistant glucose transport in adipose cells without restoring glucose transporter gene expression. J Clin Invest 1991; 87: 561-70.

[17]   List JF, Woo V, Morales E, Tang W, Fiedorek FT. Sodium-glucose cotransport inhibition with dapagliflozin in type 2 diabetes mellitus. Diabetes Care 2009; 32: 650-7.

[18]   Katsuno K, Fujimori Y, Takemura Y, *et al.,* Sergliflozin, a novel selective inhibitor of low-affinity sodium glucose cotransporter (SGLT2), validates the critical role of SGLT2 in renal glucose reabsorption and modulates plasma glucose level. J Pharmacol Exp Ther 2007; 320: 323-30.

[19]   Chassis H, Jolliffe N, Smith H. The action of phlorizin on the excretion of glucose, xylose, sucrose, creatinine, and urea by man. J Clin Invest 1933; 12: 1083-9.

[20]   Vick HD, Deidrich DF. Reevaluation of renal tubular glucose transport inhibition by phlorizin analogs. Am J Physiol 1973; 224: 552-7.

[21]   Silverman M. Glucose transport in the kidney. Biochim Biophys Acta 1976; 457: 303-51.

[22]   Ehrenkranz RRL, Lewis NG, Kahn CR, Roth J. Phlorizin: a review. Diabetes Metab Res Rev 2005; 21: 31-8.

[23]   Komoroski B, Vachharajani N, Boulton D, *et al.,* Dapagliflozin, a novel SGLT2 inhibitor, induces dose-dependent glucosuria in healthy subjects. Clin Pharmacol Ther 2009; 85: 520-6.

[24]   Komoroski B, Vachharajani N, Feng Y, Li L, Kornhauser D, Pfister M. Dapagliflozin, a novel, selective SGLT2 inhibitor, improved glycemic control over 2 weeks in patients with type 2 diabetes mellitus. Clin Pharmacol Ther 2009; 85: 513-9.

[25]   Bailey CJ, Iqbal N, T'joen C, List JF. Dapagliflozin monotherapy in drug-naïve patients with diabetes: a randomized-controlled trial of low-dose range. Diabetes Obes Metab 2012; 14: 951-9

[26]   Kaku K, Inoue S, Matsuoka O, *et al.,* Efficacy and safety of dapagliflozin as a monotherapy for type 2 diabetes mellitus in Japanese patients with inadequate glycaemic control: a Phase II multicentre, randomized, double-blind, placebo-controlled trial. Diabetes Obes Metab 2012 [Epub ahead of print]

[27]   Ferrannini E, Ramos SJ, Salsali A, Tang W, List JF. Dapagliflozin monotherapy in type 2 diabetic patients with inadequate glycemic control by diet and exercise: a randomized, double-blind, placebo-controlled, phase 3 trial. Diabetes Care 2010; 33: 2217-24.

[28]   List JF, Woo V, Morales E, Tang W, Fiedorek FT. Sodium-glucose cotransport inhibition with dapagliflozin in type 2 diabetes. Diabetes Care 2009; 32: 650-7.

[29]   Henry RR, Murray AV, Marmolejo MH, Hennicken D, Ptaszynska A, List JF. Dapagliflozin, metformin XR, or both: initial pharmacotherapy for type 2 diabetes, a randomised controlled trial. Int J Clin Pract 2012; 66: 446-56.

[30]   Bailey CJ, Gross JL, Pieters A, Bastien A, List JF. Effect of dapagliflozin in patients with type 2 diabetes who have inadequate glycaemic control with metformin: a randomised, double-blind, placebo-controlled trial. Lancet 2010; 375: 2223-33.

[31]   Nauck MA, Del Prato S, Meier JJ, *et al.,* Dapagliflozin *vs.* glipizide as add-on therapy in patients with type 2 diabetes who have inadequate glycemic control with metformin: a randomized, 52-week, double-blind, active-controlled noninferiority trial. Diabetes Care 2011; 34: 2015-22.

[32]   Strojek K, Yoon KH, Hruba V, Elze M, Langkilde AM, Parikh S. Effect of dapagliflozin in patients with type 2 diabetes who have inadequate glycaemic control with glimepiride: a randomized, 24-week, double-blind, placebo-controlled trial. Diabetes Obes Metab 2011; 13: 928-38.

[33]   Rosenstock J, Vico M, Wei L, Salsali A, List JF. Effects of dapagliflozin, an SGLT2 inhibitor, on HbA(1c), body weight, and hypoglycemia risk in patients with type 2 diabetes inadequately controlled on pioglitazone monotherapy. Diabetes Care 2012; 35: 1473-8.

[34]   Wilding JP, Woo V, Soler NG, *et al.,* Dapagliflozin 006 Study Group. Long-term efficacy of dapagliflozin in patients with type 2 diabetes mellitus receiving high doses of insulin: a randomized trial. Ann Intern Med 2012; 156: 405-15.

[35]   Wilding JP, Norwood P, T'joen C, Bastien A, List JF, Fiedorek FT. A study of dapagliflozin in patients with type 2 diabetes receiving high doses of insulin plus insulin sensitizers: applicability of a novel insulin-independent treatment. Diabetes Care 2009; 32: 1656-62.

[36]   Kasichayanula S, Liu X, Shyu WC, *et al.,* Lack of pharmacokinetic interaction between dapagliflozin, a novel sodium-glucose transporter 2 inhibitor, and metformin, pioglitazone, glimepiride or sitagliptin in healthy subjects. Diabetes Obes Metab 2011; 13: 47-54.

[37]   Sha S, Devineni D, Ghosh A, *et al.,* Canagliflozin, a novel inhibitor of sodium glucose co-transporter 2, dose dependently reduces calculated renal threshold for glucose excretion and increases urinary glucose excretion in healthy subjects. Diabetes Obes Metab 2011; 13: 669-72.

[38]   Inagaki N, Kondo K, Iwasaki T, *et al.,* Canagliflozin, a novel inhibitor of sodium glucose cotransporter 2 (SGLT2) improves glycemic control and reduces body weight in Japanese type 2 diabetes Mellitus (T2DM). Diabetes 2011; 60(suppl): A999.

[39]   Rosensotck J, Arbit D, Usiskin K, Capuano G, Canovatchel W. Canagliflozin an inhibitor of sodium glucose co-transporter 2 (SGLT2), improves glycemic control and lowers body weight in subjects with type 2 diabetes (T2D) on metformin. Diabetes 2010; 59 suppl 1: A21.

[40]   Devineni D, Morrow L, Hompesch M, *et al.,* Canagliflozin improves glycemic control over 28 days in subjects with type 2 diabetes not optimally controlled on insulin. Diabetes Obes Metab 2012; 14: 539-45.

[41]   Polidori D, Zhao Y, Sha S, Canovatchel W. Canagliflozin treatment improves beta cell function in subject with type 2 diabetes. Diabetes 2010; 59 suppl 1: A176.

[42]   Stenlöf K, Cefalu WT, Kim KA, *et al.,* Long-Term Efficacy and Safety of Canagliflozin Monotherapy in Patients With Type 2 Diabetes Inadequately Controlled With Diet and Exercise: Findings from the 52-Week CANTATA-M Study. Curr Med Res Opin. 2013 Sep 30. [Epub ahead of print]

[43]   Lavalle-González FJ, Januszewicz A, Davidson J, *et al.,* Efficacy and safety of canagliflozin compared with placebo and sitagliptin in patients with type 2 diabetes on background metformin monotherapy: a randomised trial. Diabetologia. 2013 Sep 13. [Epub ahead of print]

[44]   Cefalu WT, Leiter LA, Yoon KH, *et al.,* Efficacy and safety of canagliflozin *vs.* glimepiride in patients with type 2 diabetes inadequately controlled with metformin (CANTATA-SU): 52 week results from a randomised, double-blind, phase 3 non-inferiority trial. Lancet. 2013; 382: 941-50.

[45]   Schernthaner G, Gross JL, Rosenstock J, *et al.,* Canagliflozin Compared With Sitagliptin for Patients With Type 2 Diabetes Who Do Not Have Adequate Glycemic Control With Metformin Plus Sulfonylurea: A 52-week randomized trial. Diabetes Care. 2013; 36: 2508-15.

[46]   Rosenstock J, Aggarwal N, Polidori D, *et al.,* Canagliflozin DIA 2001 Study Group. Dose-ranging effects of canagliflozin, a sodium-glucose cotransporter 2 inhibitor, as add-on to metformin in subjects with type 2 diabetes. Diabetes Care. 2012; 35: 1232-8.

[47]   Devineni D, Morrow L, Hompesch M, *et al.,* Canagliflozin improves glycaemic control over 28 days in subjects with type 2 diabetes not optimally controlled on insulin. Diabetes Obes Metab. 2012; 14: 539-45.

[48]   Bode B, Stenlöf K, Sullivan D, Fung A, Usiskin K. Efficacy and safety of canagliflozin treatment in older subjects with type 2 diabetes mellitus: a randomized trial. Hosp Pract (1995). 2013; 41: 72-84.

[49]   Yale JF, Bakris G, Cariou B, *et al.,* Efficacy and safety of canagliflozin in subjects with type 2 diabetes and chronic kidney disease. Diabetes Obes Metab. 2013;15: 463-73.

[50]   Stenlöf K, Cefalu WT, Kim KA, *et al.,* Effect of canagliflozin, a sodium glucose co-transporter 2 (SGLT2) inhibitor, on bacteriuria and urinary tract infection in subjects with type 2 diabetes enrolled in a 12-week, phase 2 study. Curr Med Res Opin. 2012; 28:1167-71.

[51]   Nyirjesy P, Zhao Y, Ways K, Usiskin K. Evaluation of vulvovaginal symptoms and Candida colonization in women with type 2 diabetes mellitus treated with canagliflozin, a sodium glucose co-transporter 2 inhibitor. Curr Med Res Opin. 2012; 28: 1173-8.

[52]   Koiwai K, Seman L, Yamamura N, *et al.,* Safety, tolerability, pharmacokinetics and pharmacodynamics of single doses of BI 10773, a sodium-glucose co-transporter inhibitor (SGLT2), in Japanese healthy volunteers. Diabetes 2010; 59 suppl 1: 2175PO.

[53]   Heise T, Seewaldt-Becker E, Macha S, *et al.,* BI 10773, a sodium-glucose cotransporter inhibitor (SGLT-2), is safe and efficacious following 4-week treatment in patients with type 2 diabetes. Diabetes 2010; 59 (suppl)· 629P.

[54]   Rosenstock J, Jelaska A, Seman L, Pinnetti S, Hantel S, Woerle HJ. Efficacy and safety of BI 10773, a new sodium glucose cotransporter-2 (SGLT-2) inhibitor, in type 2 diabetes inadequately controlled on metformin. Diabetes 2011; 60(suppl): 989P.

[55]   Kashiwagi A, Utsuno A, Kazuta K, Yoshida S, Kageyama S. ASP1941, a novel, selective SGLT2 inhibitor, was effective and safe in Japanese healthy volunteers and patients with type 2 diabetes mellitus. Diabetes 2010; 59 suppl 1: A21.

[56]   Takinami A, Takinami Y, Kazuta K, *et al.,* Ipragliflozin improved glycemic control with additional benefit of reduction of body weight and blood pressure in Japanese patients with type 2 diabetes mellitus BRIGHTEN Study. Diabetologia 2011; 54(suppl): A149.

[57]   Fujimori Y, Katsuno K, Nakashima I, Ishikawa-Takemura Y, Fujikura H, Isaji M. Remogliflozin etabonate, in a novel category of selective low-affinity sodium glucose cotransporter (SGLT2) inhibitors, exhibits antidiabetic efficacy in rodent models. J Pharmacol Exp Ther 2008; 327: 268-76.

[58]   Hussey EK, O'Connor-Semmes RL, Tao W, Poo JL, Dobbins RL. Safety, pharmacokinetics and pharmacodynamics of remogliflozin etabonate (SGLT2 inhibitor) and metformin when co-administered in type 2 diabetes mellitus patients. 69th Scientific Session of the American Diabetes Association, New Orleans, USA (2009).

[59]   Kapur AR, Hussey E, Dobbins RL, Tao W, Hompesch M, Nunez DJ. First human dose escalation study with remogliflozin etabonate (RE) in healthy subjects and in subjects with type 2 diabetes mellitus. 69th Scientific Session of the American Diabetes Association, New Orleans, USA (2009).

[60]   Komoroski B, Vachharajani N, Feng Y, Li L, Kornhauser D, Pfister M. Dapagliflozin, a novel, selective SGLT2 inhibitor, improved glycemic control over 2 weeks in patients with type 2 diabetes mellitus. Clin Pharmacol Ther 2009; 85: 513-9.

[61]   Geerlings SE, Stolk RP, Camps MJ, Netten PM, Collet TJ, Hoepelman AI. Diabetes Women Asymptomatic Bacteriuria Utrecht Study Group.Risk factors for symptomatic urinary tract infection in women with diabetes. Diabetes Care 2000; 23: 1737-41.

[62]   http://www.fda.gov/downloads/AdvisoryCommittees/Committees MeetingMaterials/Drugs/EndocrinologicandMetabolicDrugsAdvisoryCommittee/UCM262 996.pdf This site contains the entire package for the FDA submission of dapagliflozin with a summary of the clinical efficacy and adverse events observed in phase 1, 2, and 3 of studies.

[63]   Astrazanacea-investigational compound dapagliflozin sustained glycaemic control and weight reduction. Eur Pharmaceut Rev 2011. http://www.europeanpharmaceuticalreview.com/7918/ news/industry-news/bristol-myers-squibb-and-astrazenecaannounce-investigational-compound-dapagliflozinsustained-glycemic-control-and-weight-reduction-in-studyof-type2-diabetes-patients-inadequately-controlled-withme/.Accessed on Aug 1, 2011.

[64]   Kadokura T, Ishikawa H, Nakajo I, *et al.,* The effect of renal impairment on the pharmacokinetics and urinary glucose excretion of the SGLT2 inhibitor ipragliflozin in Japanese type 2 diabetes mellitus patients. Diabetologia 2011; 54(suppl): A847.

CHAPTER 4

Osteocalcin Plays as an Endocrine Hormone in Glucose Metabolism

Ippei Kanazawa[*]

Department of Internal Medicine 1, Shimane University Faculty of Medicine, Shimane, Japan

Abstract: The number of patients with diabetes mellitus and osteoporosis are rapidly increasing especially in industrialized countries. Accumulating evidence shows that diabetes and osteoporotic fractures are associated with each other. Patients with type 2 diabetes have a higher risk of fractures independent of bone mineral density compared to healthy subjects. Several key endocrine factors in glucose metabolism such as insulin and adiponectin are shown to regulate osteoblastogenesis and bone turnover. On the other hand, an excellent study using genetic mutant mice models previously demonstrated that osteocalcin, one of the osteoblast-specific proteins, has a hormonal function. Osteocalcin increases the expression of insulin in pancreatic β-cells as well as adiponectin in adipocytes, resulting in preventing diabetes mellitus. Furthermore, the insulin signaling in osteoblasts is reported to regulate osteocalcin activity as well as peripheral adiposity and glucose metabolism. Also, several clinical studies showed that serum osteocalcin was inversely associated with blood glucose and visceral fat mass and positively with serum adiponectin levels, parameters of insulin secretion and its sensitivity in humans. Altogether, these experimental and clinical findings suggest that bone metabolism and glucose homeostasis are associated with each other through the action of osteocalcin. These findings are reviewed herein.

Keywords: Osteocalcin, insulin, adiponectin, diabetes mellitus, osteoblast, osteoclast, bone metabolism.

INTRODUCTION

The number of patients with diabetes mellitus and osteoporosis is rapidly increasing in industrialized countries. Although the incidence of osteoporosis and type 2 diabetes mellitus is known to increase in prevalence with aging, both

---

[*]**Address correspondence to Ippei Kanazawa:** Department of Internal Medicine 1, Shimane University Faculty of Medicine, 89-1 Enya-cho, Izumo 693-8501, Japan; Tel: +81-853-20-2183; Fax: +81-853-23-8650; E-mail: ippei.k@med.shimane-u.ac.jp

Atta-ur-Rahman (Ed)

diseases were traditionally viewed as separate entities. However, the relationship between diabetes and osteoporotic fractures is recently and increasingly becoming recognized. Cumulative evidence shows that there is a positive correlation between bone mineral density (BMD) and fat mass, suggesting that body fat and bone mass are related to each other [1-3]. Adipose tissue has been shown to secrete a variety of biologically active molecules [4], which are named adipokines such as adiponectin, leptin and resistin. These adipokines are specifically and highly expressed in visceral and subcutaneous fat mass. They are abundantly present in plasma and have been proposed to play important roles in the regulation of energy homeostasis and insulin sensitivity [5-7]. Previous studies including our clinical ones indicated that serum levels of these adipokines were associated with BMD, bone biochemical markers, and a risk of fractures [8-10]. These findings suggest that fat mass could affect bone metabolism by secreting the adipokines.

On the other hand, recent studies suggested that secreted proteins from bone affect glucose homeostasis and fat mass. Osteocalcin, an osteoblast-specific protein, has several hormonal features and is secreted in the general circulation from osteoblastic cells. To examine the role of osteocalcin in bone, *Osteocalcin* knockout (*Ocn$^{-/-}$*) mice were previously generated. While analyzing these mutant mice, an abnormal accumulation of visceral fat was surprisingly found [11]. Furthermore, the *Ocn$^{-/-}$* mice displayed hyperglycemia and glucose intolerance. In addition, β-cell proliferation and insulin secretion were decreased, and insulin resistance by the inhibition of adiponectin expression was observed.

Osteocalcin has 49 amino acids and undergoes γ-carboxylation of glutamyl residues at positions 17, 21 and 24, which facilitates binding of osteocalcin to hydroxyapatite in bone. Several *in vitro* and *in vivo* examinations suggest that uncarboxylated form of osteocalcin is an active form in glucose metabolism [11, 12]. In this review, I describe the role of osteocalcin in glucose metabolism and the association between serum osteocalcin level and parameters of glucose homeostasis in humans.

THE ROLE OF OSTEOCALCIN IN GLUCOSE HOMEOSTASIS

Lee *et al.* reported for the first time an excellent animal study using two gene mutant mice models showing that osteocalcin might affect systemic glucose metabolism [11]. To investigate the function of osteocalcin in adipocytes and pancreatic β-cells, $Ocn^{-/-}$ mice were examined. $Ocn^{-/-}$ mice displayed hyperglycemia and glucose intolerance, decreased β-cell and insulin secretion, decreased insulin sensitivity by the inhibition of adiponectin expression, and increased fat mass and serum triglyceride level. In addition, coculture COS cells transfected with an *Ocn* expression vector with islets or adipocytes showed increased expression of insulin and adiponectin. Furthermore, injection of recombinant osteocalcin improved the glucose intolerance and increased serum insulin level.

The expression of *Esp* which encodes osteotesticular protein tyrosine phosphatase (OST-PTP) is restricted to osteoblasts, sertoli cells and embryonic stem cells [13]. OST-PTP is a transmembrane tyrosine phosphatase which cannot directly affect distant tissues like fat and β-cells as well as stimulates carboxylation of osteocalcin and decreases osteocalcin bioactivity. Therefore, they examined the *Esp* null mice ($Esp^{-/-}$) as a model of gain of osteocalcin bioactivity. In contrast to $Ocn^{-/-}$ mice, $Esp^{-/-}$ mice displayed lower blood glucose levels of fasting and after glucose injection, increased the expression of insulin and its serum concentration, as well as increased insulin sensitivity and adiponectin expression. Moreover, the mice showed decreased fat mass and serum triglyceride level, a resistance to obesity and diabetes induced by high fat diet, as well as a resistance to diabetes induced by streptozotocin. The metabolic phenotype of $Ocn^{-/-}$ mice is the mirror image of the one observed in $Esp^{-/-}$ mice. They then examined whether metabolic abnormalities of $Esp^{-/-}$ mice could be corrected by reducing osteocalcin expression. $Esp^{-/-}$ mice were crossed with $Ocn^{+/-}$. In $Esp^{-/-}$ and $Ocn^{+/-}$ mice, the metabolic abnormalities such as hypoglycemia, hyperinsulimenia and increased serum adiponectin level were completely reversed.

Because carboxylated osteocalcin has a higher affinity for hydroxyapatite than uncarboxylated osteocalcin, it is assumed that uncarboxylated osteocalcin is an active form. It is thus examined whether OST-PTP could affect carboxylation of osteocalcin. Sera from wild-type (WT) and $Esp^{-/-}$ mice were added to hydroxyapatite beads. It was found that 90% of serum osteocalcin in WT mice was bound to hydroxyapatite, whereas only 74% was bound in the serum from $Esp^{-/-}$ mice. This finding suggests that OST-PTP influences osteocalcin function by regulating its degree of γ-carboxylation. Finally, they used carboxylated and uncarboxylated osteocalcin in cell-based assays. Uncarboxylated osteocalcin stimulated the expression of adiponectin in adipocytes as well as of cyclin D1 and insulin in islets, while carboxylated one showed no effect. Further examination was then performed to examine the effect of administration of recombinant uncarboxylated osteocalcin on glucose metabolism in WT mice [12]. Continuous injection of low doses of uncarboxylated osteocalcin increased insulin secretion, β-cell proliferation, insulin sensitivity and adiponectin expression as well as decreased fat mass in the WT mice. In addition, osteocalcin injection prevented high-fat diet- and gold thioglucose-induced obesity and diabetes. They then tested the therapeutic potential of intermittent administration of recombinant uncarboxylated osteocalcin in WT mice [14]. Daily injection of uncarboxylated osteocalcin significantly improved glucose intolerance and insulin resistance in mice fed a normal diet as well as a high-fat diet. Moreover, hepatic steatosis induced by the high-fat diet was completely rescued in mice receiving osteocalcin daily injection. These findings suggest that the administration of uncarboxylated osteocalcin might be useful for treatment of type 2 diabetes and obesity (Fig. **1**).

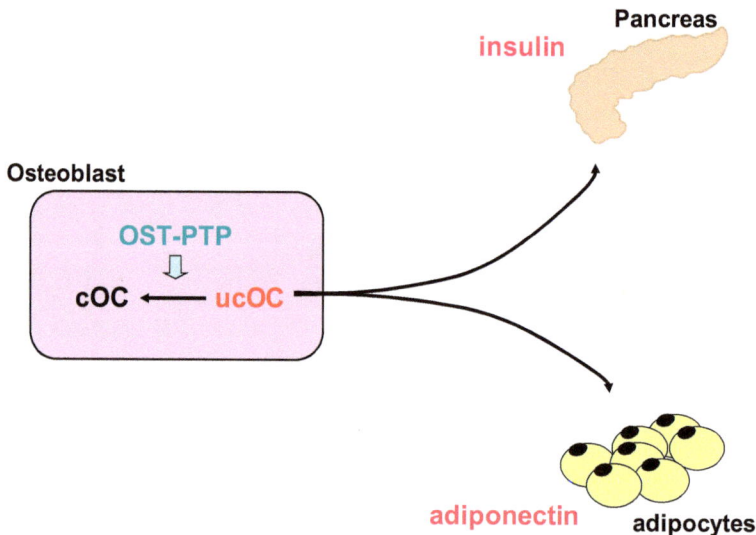

Figure 1: The regulation of glucose homeostasis by osteoblasts. Uncarboxylated osteocalcin (ucOC) secreted from osteoblast stimulates insulin expression in pancreas and adiponectin in adipocytes, regulating systemic glucose metabolism. Osteotesticular protein tyrosine phosphatase (OST-PTP), which *Esp* gene codes, accelerates carboxylation of ucOC.

OSTEOCALCIN RECEPTOR AND THE SIGNALING PATHWAY

Although previous studies showed that uncarboxylated osteocalcin affected the expression of insulin and adiponectin, its receptor and signaling pathway were unclear. It is recently reported that G-protein coupled receptor family C group 6 member A (GPRC6A) is a candidate for mediating the response to osteocalcin [15]. GPRC6A is an orphan receptor belonging to the C family of G protein-coupled receptors (GPCRs), which is known as seven-transmembrane domain receptors, and is widely expressed and sense amino acids and extracellular calcium [16, 17]. GPRC6A null mice have osteopenia, hepatic steatosis, hyperglycemia, glucose intolerance, and insulin resistance [18], suggesting that the overall function of GPRC6A may coordinate the anabolic responses of multiple tissues. According to the metabolic abnormalities of the GPRC6A null mice, Pi *et al.* hypothesized that GPRC6A might participate in the function of osteocalcin in glucose metabolism. Therefore, they examined the effect of osteocalcin on the cells expressing GPRC6A. Recombinant human osteocalcin stimulated ERK activity in HEK-293 cells overexpressing *Gprc6a* in a dose-dependent manner, but not in untransfected controls. They also found that a cell line of mouse pancreatic β-cell cells and pancreas isolated from WT mice expressed *Gprc6a*, and that osteocalcin treatment resulted in a dose-dependent stimulation of ERK in the cells *in vitro* and in pancreas *in vivo*. In addition, administration of recombinant human osteocalcin induced a 3-fold increase in insulin expression in pancreas as well as a 40% increase in serum insulin level in WT mice, but not in GPRC6A null mice. Thus, these results suggest that GPRC6A is a biologically relevant osteocalcin-sensing receptor. However, the authors did not mention whether the recombinant osteocalcin they used in the study is uncarboxylated or carboxylated one. Furthermore, it was not examined whether or not the effects of osteocalcin are directly and solely mediated by GPRC6A. They also found that the response of GPRC6A to osteocalcin was similar to that of calcium and arginine which are known as GPRC6A ligands. It is thus speculated that GPRC6A could sense both nutrient derived factors, such as calcium and amino acids, as well as osteocalcin. Finally, GPRC6A is widely expressed in multiple tissues and loss of *Gprc6a* results in multiple metabolic abnormalities. Additional examination using conditional deletion of *Gprc6a* in specific tissue will be necessary to establish the tissue-specific functions of GPRC6A. Most recently, another research group showed that osteocalcin binds to GPRC6A presented in testis by using specific antibody for osteocalcin [19], suggesting that osteocalcin acts directly *via* GPRC6A. As described above, previous studies demonstrated that uncarboxylated osteocalcin, but carboxylated, is an active form in glucose homeostasis. However, it is not reported in the paper that the antibody is specific for uncarboxylated osteocalcin or not. Thus, it is still unclear GPRC6A can specifically drive the uncarboxylated osteocalcin signaling.

ENDOCRINE LOOPS AMONG BONE, PANCREAS, AND ADIPOSE TISSUE

Type 1 diabetes mellitus is caused by autoimmune destruction of insulin-producing pancreatic β-cells. It has been shown that patients with type 1 diabetes have osteopenia and osteoporosis with reduced bone formation [20, 21] and an increased risk of fragility fractures [22]. Previous studies have shown that osteoblast has a functional insulin receptor and that insulin stimulates the proliferation and differentiation of osteoblasts as well as collagen synthesis [23, 24]. It has been shown that deletion of insulin receptor in osteoblasts induced decreases in alkaline phosphatase (ALP) activity and osteocalcin expression by suppressing a Runx2 inhibitor, Twist2 [25]. Moreover, osteoblast-specific insulin receptor knockout ($IR_{osb}{}^{-/-}$) mice displayed reduced bone accumulation due to decreased bone formation and deficient numbers of osteoblasts. These findings indicate that insulin signaling in osteoblasts has a pivotal role in bone formation and bone development.

With regard to the hormonal loop networks, it is rational to suggest that signals derived from the osteoblast might regulate insulin expression and secretion in pancreas. Recently, two independent groups revealed that insulin receptor signaling in osteoblasts regulated peripheral adiposity and glucose metabolism [25, 26]. Fulzele *et al.* demonstrated that $IR_{osb}{}^{-/-}$ mice driven by *Osteocalcin-Cre* developed marked peripheral adiposity and hyperglycemia accompanied by severe glucose intolerance and insulin resistance [25]. The weight of total body and fat mass were significantly greater in the $IR_{osb}{}^{-/-}$ mice than those in WT mice, and measurements of body composition revealed a remarkable increase in fat mass and a decrease in lean mass in the $IR_{osb}^{-/-}$ mice. In addition, $IR_{osb}^{-/-}$ had decreased rates of oxygen consumption and energy expenditure compared to controls. Serum glucose after glucose injection was significantly higher in $IR_{osb}^{-/-}$ mice than WT mice, while serum insulin level was decreased in $IR_{osb}^{-/-}$ mice. Furthermore, β-cell mass and insulin expression were decreased in $IR_{osb}^{-/-}$ mice, and insulin tolerance test and gene expression analysis in adipose tissue indicated that $IR_{osb}^{-/-}$ mice had a severe insulin resistant with low adiponectin expression. They found that serum uncarboxylated form of osteocalcin was decreased in $IR_{osb}^{-/-}$ mice, and that infusion of uncarboxylated osteocalcin reversed their glucose intolerance. Ferron *et al.* also showed that analysis of $IR_{osb}^{-/-}$ mice by crossing *collagen1α1-Cre* transgenic mice showed that insulin signaling in osteoblasts contributes to whole-body glucose homeostasis by increasing β-cell proliferation, insulin secretion, and energy expenditure [26]. Taken together, these findings indicate the existence of a bone-pancreas endocrine feedforward loop through insulin signaling in the osteoblasts and ensure that insulin signaling in osteoblasts stimulates the differentiation of the cells and osteocalcin production, which in turn regulates insulin sensitivity and pancreatic insulin secretion.

Obesity and fat mass accumulation are known to be associated with bone metabolism. Previous studies on adipocyte function have revealed that not only is adipose tissue an energy-storing organ but also it secretes a variety of biologically active molecules, which are named adipokines [4]. It has been shown that osteoblast has an adiponectin receptor and the adiponectin signaling stimulates the differentiation of osteoblasts as well as osteocalcin expression. Luo *et al.* reported that recombinant adiponectin increased ALP activity and osteocalcin expression in human osteoblasts [27]. Previously, we also demonstrated that a knockdown of adiponectin receptor expression by siRNA induced an inhibition of ALP activity and osteocalcin expression in osteoblastic MC3T3-E1 cells [28]. In addition, adiponectin is reported to stimulate the differentiation of osteoclasts *via* increasing receptor activator of nuclear factor kappa-B ligand (RANKL) expression and decreasing Osteoprotegerin (OPG) expression in osteoblasts although adiponectin did not affect osteoclasts directly [29]. Since adiponectin stimulates the expression of osteocalcin as well as the differentiation of osteoclasts and osteocalcin alternatively stimulates the expression of adiponectin in adipocytes, it is rational to hypothesize that another endocrine loop might exist between bone and adipose tissue through adiponectin and osteocalcin (Fig. **2**).

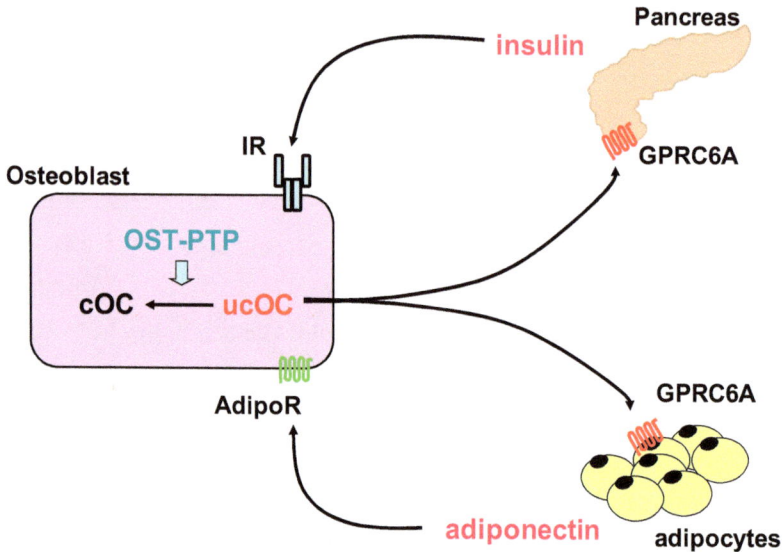

Figure 2: Relationships among bone, pancreas, and adipose tissue. Adiponectin as well as insulin stimulate the differentiation of osteoblasts and osteocalcin expression in osteoblasts through adiponectin receptor (AdipoR) and insulin receptor (IR). Uncarboxylated osteocalcin (ucOC) secreted from osteoblasts alternatively stimulates the expression of insulin in β-cells and of adiponectin in adipocytes *via* G-protein coupled receptor family C group 6 member A (GPRC6A).

ATF4 AND FOXO1 IN OSTEOBLASTS REGULATE GLUCOSE METABOLISM THROUGH THE OSTEOCALCIN FUNCTION

Many transcriptional factors are involved in the osteoblast differentiation and function. Runx2, Osterix, and activating transcription factor 4 (ATF4) are reported to accumulate only or mostly in osteoblasts [30-33]. Runx2 and Osterix play critical roles in the commitment of mesenchymal stem cells into the osteoblast linage, while ATF4, a member of the cAMP-responsive element-binding protein (CREB) family of basic zipper-containing proteins, regulates the terminal differentiation of osteoblasts and bone mineralization [33]. In addition, ATF4 promotes bone resorption through increasing the expression of RANKL in osteoblasts [34]. Since ATF4 stimulates the expression of osteocalcin and bone turnover, it was assumed that ATF4 might be involved in the glucose metabolism through the uncarboxylated osteocalcin function. However, it was surprisingly found that *Atf4* knockout (*Atf4*$^{-/-}$) mice displayed significant lower fat mass and blood glucose level compared to their littermate controls [35]. In the *Atf4*$^{-/-}$ mice, insulin expression and β-cell mass in pancreas as well as insulin sensitivity in liver, fat, and muscle were increased. In contrast, *Atf4* overexpression in osteoblasts showed hyperglycemia, lower serum insulin level, less β-cell mass, and insulin resistance compared to WT mice. Moreover, the metabolic phenotype of *Atf4*$^{-/-}$ mice was completely reversed by *Atf4* transgene in osteoblasts. These findings suggest that ATF4 inhibits insulin secretion and decreases insulin sensitivity although ATF4 increases osteocalcin expression. They then focused on the effect of ATF4 on *Esp* expression, which inhibits the osteocalcin bioactivity, and found that ATF4 increased *Esp* expression in osteoblasts. Indeed, the uncarboxylated form of osteocalcin was decreased in *Atf4* overexpression in osteoblast, while that was increased in *Atf4*$^{-/-}$ mice. It was thus concluded that ATF4 might control the expression of osteocalcin and to a larger extent of *Esp*, causes a chronic decrease in osteocalcin-mediated insulin production and induces a long-term negative effect on pancreatic β-cell proliferation.

Insulin signaling pathway has been studied intensively since insulin resistance plays a pivotal role in the pathogenesis of metabolic diseases including type 2 diabetes, hypertension, dyslipidemia, and atherosclerosis. FoxO1, which belongs to the Forkhead family of transcription factors, is discovered as the major transcriptional mediator of insulin signaling and insulin transmits its signal by inhibiting FoxO1 activity in β-cell, adipocyte, hepatocyte, and myoblast [36]. Although FoxO1 is expressed in many cells including osteoblasts, its functions were still unclear especially in osteoblast. Rached *et al.* generated osteoblast-specific FoxO1 knockout (*FoxO1*$_{osb}$$^{-/-}$) mice and examined the role of FoxO1 in osteoblasts on systemic glucose metabolism [37]. Marked reduction of bone mass was observed in the *FoxO1*$_{osb}$$^{-/-}$ mice, and Luciferase assay using osteocalcin promoter gene indicated that interaction of FoxO1 with ATF4 stimulated osteocalcin expression while FoxO1 alone suppressed it [38]. The *FoxO1*$_{osb}$$^{-/-}$ mice showed hypoglycemia due to hyperinsulinemia with an increase in β-cell mass. In addition, gonadal fat pad and liver fat content were decreased while serum adiponectin and insulin sensitivity were increased in the *FoxO1*$_{osb}$$^{-/-}$ mice. The *FoxO1*$_{osb}$$^{-/-}$ mice have a phenotype that mirrors the metabolic phenotype of mice lacking osteocalcin, thus suggesting a gain of osteocalcin activity in $FoxO1_{osb}^{-/-}$ mice. Indeed, deletion of a single osteocalcin allele from $FoxO1_{osb}^{-/-}$ mice ($FoxO1_{osb}^{-/-};Ocn^{+/-}$) resulted in a complete reversal of the metabolic abnormalities of the $FoxO1_{osb}^{-/-}$ mice. Furthermore, hydroxyapatite binding assay indicated that 37% of osteocalcin present in the serum of $FoxO1_{osb}^{-/-}$ mice was uncarboxylated, whereas only 27% of uncarboxylated osteocalcin was present in the serum of WT mice. These experiments suggest that FoxO1 in osteoblasts controls glucose homeostasis by regulating both the expression and carboxylation of osteocalcin. Based on these results, it was examined whether FoxO1 acts on downstream of the insulin signaling pathway in osteoblasts from $IR_{osb}^{-/-}$ and $FoxO1_{osb}^{+/-}$ mice, and found that decreasing FoxO1 expression corrected the glucose intolerance of $IR_{osb}^{-/-}$ mice [26]. Taken together, these findings indicate that insulin signaling in osteoblasts affects glucose homeostasis by inhibiting the FoxO1 activity.

DECARBOXYLATION OF OSTEOCALCIN IN BONE EXTRA-CELLULAR MATRIX BY OSTEOCLASTS IS REQUIRED TO REGULATE GLUCOSE METABOLISM

Recently, it was found that not only osteocalcin production in osteoblasts but also osteoclast activity is important for the function of osteocalcin. Ferron *et al.* noticed a marked decrease in serum level of CTx, a marker of bone resorption, in the $IR_{obs}^{-/-}$ mice [26]. Coculture analysis with osteoclast precursor cells and $IR^{-/-}$ osteoblasts isolated from the knockout mice showed that the area covered by resorption pits was significantly decreased. They found that expression of *Opg*, a gene encoding a decoy receptor for RANKL and a negative regulator of osteoclast differentiation, was increased in $IR^{-/-}$ osteoblasts, and that insulin treatment decreased *Opg* expression and secretion in WT osteoblasts, but not in $IR^{-/-}$ osteoblasts. Moreover, expression of *CathepsinK* (*Ctsk*) and *Tcirg1*, two genes implicated in bone resorption, was decreased in $IR_{obs}^{-/-}$ bone. The expression of *Ctsk* and *Tcirg1* was also decreased in osteoclasts in the coculture of osteoclast precursor cells and $IR^{-/-}$ osteoblasts. Taken together, insulin signaling in osteoblasts favors the activation of osteoclasts and bone resorption by inhibiting *Opg* expression.

It is previously reported that *Tcirg1* encodes a vacuolar proton pump subunit essential for acidification of bone extracellular matrix (ECM), and that acidification of the bone ECM is a perquisite for bone resorption [39]. Since acid pH can decarboxylate proteins [40], it was assumed that the insulin signal in osteoblasts regulates glucose homeostasis by stimulating osteoclast activity followed by decarboxylation of osteocalcin. *Oc/oc* mice, a model of loss-of-function in *Tcirg1*, result in osteopetrosis [41]. In the *oc/oc* mice, 30% decrease in the level of uncarboxylated osteocalcin as well as glucose intolerance with a marked decrease in serum insulin level, pancreas insulin content, and insulin expression in pancreas were observed [26]. To test whether the regulation of glucose metabolism by insulin signaling in osteoblasts depends on the osteoclasts ability, $IR_{obs}^{+/-};oc/+$ mice were then generated. Although $IR_{obs}^{+/-}$ or $oc/+$ mice did not show glucose intolerance, $IR_{obs}^{+/-};oc/+$ mice displayed a significant decrease in insulin secretion and glucose tolerance. $Esp^{-/-}$ mice showed higher uncarboxylated osteocalcin, lower blood glucose level, as well as increased insulin expression and sensitivity, whereas $Esp^{-/-};oc/+$ mice had normal bone resorption, normal osteocalcin carboxylation status, normal insulin secretion, normal glucose, and insulin tolerance. In addition, when the $Esp^{-/-}$ mice were treated with alendronate, a bisphosphonate, osteocalcin carboxylation, insulin secretion, blood glucose, and insulin tolerance were normalized. Finally, RANKL treatment induced bone resorption and increased serum level of uncarboxylated osteocalcin, resulting in decreases in blood glucose levels, insulin resistance, and fat mass in WT mice fed a high-fat diet than control mice. Taken together, these experiments indicate that bone resorption is necessary and sufficient to activate osteocalcin in ECM and influence glucose homeostasis (Fig. **3**).

THE EVIDENCE OF THE ASSOCIATION BETWEEN UNDER-CAR-BOXYLATED OSTEOCALCIN AND GLUCOSE METABOLISM IN HUMANS

Since the new concept of endocrine loops between bone, pancreas, and adipose tissue was reported, of particular interest is whether osteocalcin level in the circulation is associated with glucose metabolism in humans. We previously reported that serum osteocalcin was inversely associated with glucose and visceral

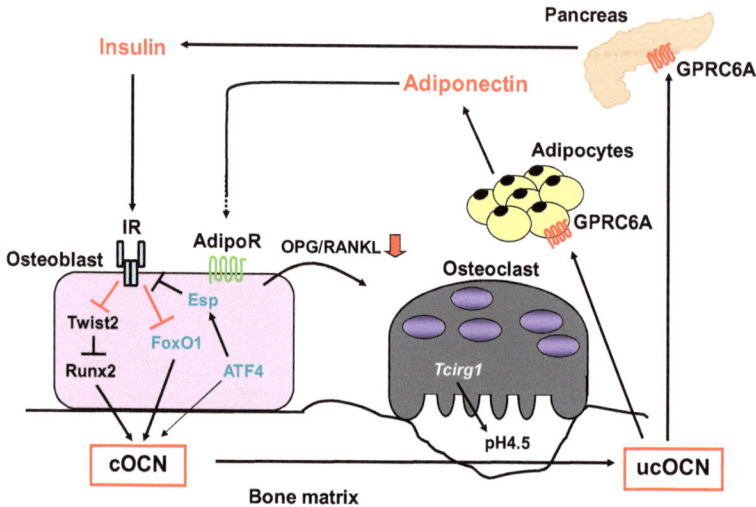

Figure 3: Bone remodeling stimulates osteocalcin activity and regulates glucose metabolism. Insulin stimulates the expression of osteocalcin by inhibiting Twist2 expression, an inhibitor of Runx2, and FoxO1 activity *via* insulin receptor (IR) in osteoblasts. Adiponectin also stimulates osteocalcin expression *via* adiponectin receptor (AdipoR). Although ATF4 stimulates osteocalcin expression, it enhances *Esp* expression more likely, resulting in an increase uncarboxylated form of osteocalcin. Insulin and adiponectin signalings decrease the ratio of osteoprotegerin (OPG) to receptor activator of nuclear factor kappa-B ligand (RANKL) and induce the differentiation of osteoclasts, resulting in decarboxylation of osteocalcin in bone extracellular matrix. Uncarboxylated osteocalcin (ucOC) stimulates the expression of insulin in β-cells and of adiponectin in adipocytes *via* G-protein coupled receptor family C group 6 member A (GPRC6A).

fat mass and positively with serum adiponectin level, parameters of insulin secretion and its sensitivity in patients with type 2 diabetes [42, 43]. Kindblom *et al.* showed that osteocalcin level was inversely related to plasma glucose level and fat mass in elderly non-diabetic persons [44]. Fernandez-Real *et al.* reported that serum osteocalcin level was associated with insulin sensitivity in non-diabetes subjects [45]. Moreover, Pittas *et al.* demonstrated that serum osteocalcin concentration was inversely associated with fasting plasma glucose, fasting insulin, a parameter of insulin resistance [homeostasis model assessment for insulin resistance (HOMA-IR)], and body fat in cross-sectional analyses [46]. They also found that osteocalcin level was associated with change in fasting plasma glucose in prospective analyses. In addition, we recently reported a longitudinal study showing that change in osteocalcin was negatively correlated with change in HbA1c during treatments of type 2 diabetes [47]. Although there were difficulties of measuring undercarboxylated osteocalcin before, a new method using monoclonal antibody, electrochemiluminescence immunoassay (ECLIA), has been developed. We measured undercarboxylated osteocalcin by the ECLIA kit and analyzed the association between undercarboxylated osteocalcin and parameters of glucose metabolism in diabetic patients. It was found for the first time that undercarboxylated osteocalcin negatively correlated with %Trunk fat and visceral/subcutaneous fat ratio as well as fasting plasma glucose and HbA1c independent of age, duration of diabetes, body stature, and renal function as well as glucose or fat metabolism, whereas bone specific ALP, another bone formation marker, did not correlate with any variable [48]. Hwang *et al.* also reported that elevated levels of undercarboxylated osteocalcin were associated with improved glucose tolerance and that undercarboxylated osteocalcin was associated with enhanced β-cell function in middle-age male healthy subjects [49]. Iki *et al.* reported that undercarboxylated osteocalcin inversely correlated with fasting plasma glucose, HbA1c and HOMA-IR even after adjustment for total osteocalcin, whereas total osteocalcin was not associated with these parameters after adjusting for undercarboxylated osteocalcin [50]. These clinical findings support the recent evidence provided by *in vivo* and *in vitro* studies that uncarboxylated/undercarboxylated osteocalcin derived from bone is involved in systemic glucose homeostasis.

As previously described, Ferron *et al.* reported that osteoclast-mediated decarboxylation of osteocalcin might be involved in osteocalcin activity and glucose metabolism. We previously reported clinical studies that a bone resorption marker was associated with glucose and fat metabolism. Urinary N-terminal cross-linked telopeptide of type-I collagen (uNTX) was independently and inversely associated with %Trunk fat, visceral fat area, visceral/subcutaneous fat ratio, and HOMA-IR as well as positively with serum adiponectin level in patients with diabetes [43, 48]. We also found that uNTX was positively associated with serum undercarboxylated osteocalcin level independent for age, height, body weight, and vitamin K intake in healthy women [51]. These findings suggest that bone resorption might be involved in the decarboxylation of osteocalcin and glucose metabolism.

Based on the evidence, it is assumed that treatments for osteoporosis may affect glucose metabolism *via* alteration of blood osteocalcin level. Because vitamin K stimulates carboxylation of osteocalcin and decreases undercarboxylated osteocalcin levels, vitamin K intake and administration might deteriorate glucose metabolism. However, there are no reports showing the negative effects of vitamin K on glucose metabolism so far. On the contrary, several studies showed that vitamin K has a benefit for glucose homeostasis. It is previously demonstrated that high vitamin K intake was associated with high insulin secretion and low insulin resistance in healthy male subjects [52, 53], and administration of vitamin K2 increased insulin sensitivity [54, 55]. Choi *et al.* recently reported a placebo-controlled trial showing that vitamin K2 administration increased a parameter of insulin sensitivity, which was associated with increased carboxylated osteocalcin, not decreased undercarboxylated one [56]. On the other hand, antiresorptive drugs for osteoporosis such as selective estrogen receptor modulator and bisphosphonate also may affect glucose metabolism *via* decreasing bone turnover and osteocalcin expression. A clinical study with type 1 diabetic patients showed that alendronate reduced the daily consumption of insulin in the patients [57]. Recently, Vestergaad *et al.* reported that subjects treated with alendronate, etidronate, and raloxifene, which all are antiresorptive drugs, had a significant lower risk of newly diagnosed type 2 diabetes [58]. These findings suggest that vitamin K and antiresorptive drugs may have protective effects on diabetes although the mechanism is still unknown, and are opposite results to cell culture and animal experiments. Since most clinically available drugs for osteoporosis target on inhibiting osteoclast activity, further studies on their effect on glucose metabolism are necessary in future. Recently, osteocalcin is reported to be expressed in not only osteoblasts but also adipocytes [59]. The expression of osteocalcin in osteoblasts might have compensatory effects on decreased serum undercarboxylated osteocalcin level by these interventions.

On the contrary, a study using intermittent administration of parathyroid hormone (PTH) 1-84, which increases bone formation and osteocalcin secretion, showed that increased undercarboxylated osteocalcin level was associated with loss of body weight and fat mass as well as an increase in serum adiponectin level after 12 months treatment [60]. This suggests that increasing osteocalcin may have a beneficial effect on glucose metabolism. Thus, further studies are needed to confirm the effect of treatments stimulating bone formation on systemic glucose homeostasis.

CONCLUSION

Emerging evidence from *in vitro* and *in vivo* experimental studies has shown that osteocalcin, especially uncarboxylated form, is a bioactive hormone, which is involved in energy homeostasis by stimulating insulin secretion in pancreas as well as adiponectin secretion in adipocytes. Therefore, the interactions exist among bone, pancreas, and adipose tissue. Moreover, several epidemiological studies support the evidence that there is the association between serum osteocalcin and glucose/fat metabolism in human. This could further be explored in the pathogenesis and therapy of metabolic diseases. However, vitamin K and antiresorptive treatments seem not to be harmful to glucose metabolism, whereas intervention with stimulating bone formation by PTH may be useful for improving glucose homeostasis. Although osteocalcin was considered as a specific protein expressed in bone, a recent study showed that it is also presented in adipocytes. There are no reports examining the role of osteocalcin expressed in adipose tissue so far. Thus, further studies are necessary to understand the role of osteocalcin in systemic glucose homeostasis.

ACKNOWLEDGEMENTS

Declared none.

CONFLICT OF INTEREST

The author confirms that this chapter content has no conflict of interest.

REFERENCES

[1]   Lim S, Joung H, Shin CS, *et al.,* Body composition changes with age have gender-specific impacts on bone mineral density. Bone 2004; 35: 792-8.

[2]   Felson DT, Zhang Y, Hannan MT, Anderson JJ. Effects of weight, and body mass index on bone mineral density in men and women. J Bone Miner Res 1993; 8: 567-73.

[3]   Glauber HS, Vollmer WM, Nevitt MC, Ensrud KE, Orwoll ES. Body weight *versus* body fat distribution, adiposity, and frame size as predictors of bone density. J Clin Endocrinol Metab 1995; 80: 1118-23.

[4]   Maeda K, Okubo K, Shimomura I, Mizuno K, Matsuzawa Y, Matsubara K. Analysis of an expression profile of genes in the human adipose tissue. Gene 1997; 190: 227-35.

[5]   Berg AH, Combs TP, Du X, Brownlee M, Scherer PE. The adipocyte-secreted protein Acrp30 enhances hepatic insulin action. Nat Med 2001; 7: 947-53.

[6]   Lee YH, Magkos F, Mantzoros CS, Kang ES. Effects of leptin and adiponectin on pancreatic β-cell function. Metabolism 2001; 60: 1664-72.

[7]   Schwartz DR, Lazar MA. Human resistin: found in translation from mouse to man. Trends Endocrinol Metab 2011; 22: 259-65.

[8]   Kanazawa I, Yamaguchi T, Yamamoto M, Yamauchi M, Yano S, Sugimoto T. Relationships between serum adiponectin levels *versus* bone mineral density, bone metabolic markers, and vertebral fractures in type 2 diabetes mellitus. Eur J Endocrinol 2009; 160: 265-73.

[9]   Yamauchi M, Sugimoto T, Yamaguchi T, *et al.,* Plasma leptin concentrations are associated with bone mineral density and the presence of vertebral fractures in postmenopausal women. Clin Endocrinol 2001; 55: 341-7.

[10]   Oh KW, Lee WY, Rhee EJ, *et al.,* The relationship between serum resistin, leptin, adiponectin, ghrelin levels and bone mineral density in middle-aged men. Clin Endocrinol 2005; 63: 131-8.

[11]   Lee NK, Sowa H, Hinoi E, *et al.,* Endocrine regulation of energy metabolism by the skeleton. Cell 2007; 130:456-69.

[12]   Ferron M, Hinoi E, Karsenty G, Ducy P. Osteocalcin differentially regulates beta cell and adipocyte gene expression and affects the development of metabolic diseases in wild-type mice. Proc Natl Acad Sci USA 2008; 105:5266-70.

[13]   Mauro LJ, Olmsted EA, Skrobacz BM, Mourey RJ, Davis AR, Dixon JE. Identification of a hormonally regulated protein tyrosine phosphatase associated with bone and testicular differentiation. J Biol Chem 1994; 269: 30659-67.

[14]   Ferron M, McKee MD, Levine RL, Ducy P, Karsenty G. Intermittent injections of osteocalcin improve glucose metabolism and prevent type 2 diabetes in mice. Bone 2012; 50: 568-75.

[15]   Pi M, Wu Y, Quarles LD. GPRC6A mediates responses to osteocalcin in β-cells *in vitro* and pancreas *in vivo*. J Bone Miner Res 2011; 26: 1680-3.

[16]   Kuang D, Yao Y, Lam J, Tsushima RG, Hampson DR. Cloning and characterization of a family C orphan G-protein coupled receptor. J Neurochem 2005; 93: 383-91.

[17]   Wellendorph P, Brauner-Osborne H. Molecular cloning, expression, and sequence analysis of GPRC6A, a novel family C G-protein-coupled receptor. Gene 2004; 335: 37-46.

[18]   Pi M, Faber P, Ekema G, *et al.,* Identification of a novel extracellular cation-sensing G-protein-coupled receptor. J Biol Chem 2005; 280: 40201-9.

[19]   Oury F, Sumara G, Sumara O, *et al.,* Endocrine regulation of male fertility by the skeleton. Cell 2011; 144: 796-809.

[20]   Kemink SA, Hermus AR, Swinkels LM, Lutterman JA, Smals AG. Osteopenia in insulin-dependent diabetes mellitus; prevalence and aspects of pathophysiology. J Endocrinol Invest 2000; 23: 295-303.

[21]   Verhaeghe J, Suiker AM, Visser WJ, Van Herck E, Van Bree R, Bouillon R. The effects of systemic insulin, insulin-like growth factor-I and growth hormone on bone growth and turnover in spontaneously diabetic BB rats. J Endocrinol 1992; 134: 485-92.

[22]   Janghorbani M, Feskanich D, Willett WC, Hu F. Prospective study of diabetes and risk of hip fracture: the Nurses' Health Study. Diabetes Care 2006; 29: 1573-8.

[23]   Pun KK, Lau P, Ho PW. The characterization, regulation, and function of insulin receptors on osteoblast-like clonal osteosarcoma cell line. J Bone Miner Res 1989; 4: 853-62.

[24]   Kream BE, Smith MD, Canalis E, Raisz LG. Characterization of the effect of insulin on collagen synthesis in fetal rat bone. Endocrinology 1985; 116: 296-302.

[25]   Fulzele K, Riddle RC, Digirolamo DJ, *et al.,* Insulin receptor signaling in osteoblasts regulates postnatal bone acquisition and body composition. Cell 2010; 142: 309-19.

[26]   Ferron M, Wei J, Yoshizawa T, *et al.,* Insulin signaling in osteoblasts integrates bone remodeling and energy metabolism. Cell 2010; 142: 296-308.

[27]   Luo XH, Guo LJ, Yuan LQ, *et al.,* Adiponectin stimulates human osteoblasts proliferation and differentiation *via* the MAPK signaling pathway. Exp Cell Res 2005; 309: 99-109.

[28]   Kanazawa I, Yamaguchi T, Yano S, Yamauchi M, Yamamoto M, Sugimoto T. Adiponectin and AMP kinase activator stimulate proliferation, differentiation, and mineralization of osteoblastic MC3T3-E1 cells. BMC Cell Biol 2007; 8: 51.

[29]   Luo XH, Guo LJ, Xie H, *et al.,* Adiponectin stimulates RANKL and inhibits OPG expression in human osteoblasts through the MAPK signaling pathway. J Bone Miner Res 2006; 21: 1648-56.

[30]   Ducy P, Zhang R, Geoffrey V, Ridall AL, Karsenty G. Osf2/Cbfa1: a transcriptional activator of osteoblast differentiation. Cell 1997; 89: 747-54.

[31]   Nakashima K, Zhou X, Kunkel G, *et al.,* The novel zinc finger-containing transcription factor osterix is required for osteoblast differentiation and bone formation. Cell 2002; 108: 17-29.

[32]   Yang X, Matsuda K, Bialek P, *et al.,* ATF4 is a substrate of RSK2 and an essential regulator of osteoblast biology; implication for Coffin-Lowry Syndrome. Cell 2004; 117: 387-98.

[33]   Yang X, Karsenty G. ATF4, the osteoblast accumulation of which is determined posttranslationally, can induce osteoblast-specific gene expression in non-osteoblastic cells. J Biol Chem 2004; 279: 47109-14.

[34]   Teitelbaum SL, Ross FP. Genetic regulation of osteoclast development and function. Nat Rev Genet 2003; 4: 638-79.

[35]   Yoshizawa T, Hinoi E, Jung DY, *et al.,* The transcription factor ATF4 regulates glucose metabolism in mice through its expression in osteoblasts. J Clin Invest 2009; 119: 2807-17.

[36]   Gross DN, Wan M, Birnbaum MJ. The role of FOXO in the regulation of metabolism. Curr Diab Rep 2009; 9: 208-14.

[37]   Rached MT, Kode A, Silva BC, *et al.,* FoxO1 expression in osteoblasts regulates glucose homeostasis through regulation of osteocalcin in mice. J Clin Invest 2010; 120: 357-68.

[38]   Rached MT, Kode A, Xu L, Yoshikawa Y, Paik JH, DePinho RA. FoxO1 is a positive regulator of bone formation by favoring protein synthesis and resistance to oxidative stress in osteoblasts. Cell Metab 2010; 11: 147-60.

[39]   Teitelbaum SL, Ross FP. Genetic regulation of osteoclast development and function. Nat Rev Genet 2003; 4: 638-49.

[40]   Engelke JA, Hale JE, Suttie JW, Price PA. Vitamin K-dependent carboxylase: utilization of decarboxylated bone Gla protein and matrix Gla protein as substrates. Biochim Biophys Acta 1991; 1078: 31-4.

[41]   Scimeca JC, Franchi A, Trojani C, *et al.,* The gene encoding the mouse homologue of the human osteoclast-specific 116-kDa V-ATPase subunit bears a deletion in osteoclastic (oc/oc) mutants. Bone 2000; 26: 207-13.

[42]   Kanazawa I, Yamaguchi T, Yamamoto M, *et al.,* Serum osteocalcin level is associated with glucose metabolism and atherosclerosis parameters in type 2 diabetes mellitus. J Clin Endocrinol Metab 2009; 94: 45-9.

[43]   Kanazawa I, Yamaguchi T, Tada Y, Yamauchi M, Yano S, Sugimoto T. Serum osteocalcin level is positively associated with insulin sensitivity and secretion in patients with type 2 diabetes. Bone 2011; 48: 270-5.

[44]   Kindblom JM, Ohlsson C, Ljunggren O, *et al.,* Plasma osteocalcin is inversely related to fat mass and plasma glucose in elderly Swedish men. J Bone Miner Res 2009; 24: 785-91.

[45]   Fernandez-Real JM, Izquierdo M, Ortega F, *et al.,* The relationship of serum osteocalcin concentration to insulin secretion, sensitivity, and disposal with hypocaloric diet and resistance training. J Clin Endocrinol Metab 2009; 94: 237-45.

[46]   Pitass AG, Harris SS, Eliades M, Stark P, Dawson-Hughes B. Association between serum osteocalcin and markers of metabolic phenotype. J Clin Endocrinol Metab 2009; 94: 827-32.

[47]   Kanazawa I, Yamaguchi T, Sugimoto T. Relationship between bone biochemical markers *versus* glucose/lipid metabolism and atherosclerosis; a longitudinal study in type 2 diabetes mellitus. Diabetes Res Clin Pract 2011; 92: 393-9.

[48]   Kanazawa I, Yamaguchi T, Yamauchi M, *et al.,* Serum undercarboxylated osteocalcin was inversely associated with plasma glucose level and fat mass in type 2 diabetes mellitus. Osteoporos Int 2011; 22: 187-94.

[49]   Hwang YC, Jeong IK, Ahn KJ, Chung HY. The uncarboxylated form of osteocalcin is associated with improved glucose tolerance and enhanced β-cell function in middle-aged male subjects. Diabetes Metab Res Rev 2009; 25: 768-72.

[50]   Iki M, Tamaki J, Fujita Y, *et al.,* Serum undercarboxylated osteocalcin levels are inversely associated with glycemic status and insulin resistance in an elderly Japanese male population: Fujiwara-kyo Osteoporosis Risk in Men (FORMEN) Study. Osteoporos Int 2012; 23: 761-70.

[51]   Yamauchi M, Yamaguchi T, Nawata K, Takaoka S, Sugimoto T. Relationships between undercarboxylated osteocalcin and vitamin K intakes, bone turnover, and bone mineral density in healthy women. Clin Nutr 2010; 29: 761-5.

[52]   Sakamoto N, Nishiike T, Iguchi H, Sakamoto K. Relationship between acute insulin response and vitamin K intake in healthy young male volunteers. Diabetes Nutr Metab 1999; 12: 37-41.

[53]   Yoshida M, Booth SL, Meigs JB, Saltzman E, Jacques PF. Phylloquinone intake, insulin sensitivity, and glycemic status in men and women. Am J Clin Nutr 2008; 88: 210-5.

[54]   Sakamoto N, Nishiike T, Iguchi H, Sakamoto K. Possible effects of one week vitamin K (menaquinone-4) tablets intake on glucose tolerance in healthy young male volunteers with different descarboxy prothrombin levels. Clin Nutr 2000; 19: 259-63.

[55]   Yoshida M, Jacques PF, Meigs JB, *et al.,* of vitamin K supplementation on insulin resistance in older men and women. Diabetes Care 2008; 31: 2092-6.

[56]   Choi HJ, Yu J, Choi H, *et al.,* Vitamin K2 supplementation improves insulin sensitivity *via* osteocalcin metabolism: a placebo-controlled trial. Diabetes Care 2011; 34: e147.

[57]   Maugeri D, Panebianco P, Rosso D, *et al.,* Alendronate reduces the daily consumption of insulin (DCI) in patients with senile type 1 diabetes and osteoporosis. Arch Gerontol Geriatr 2002; 34: 117-22.

[58]   Vestergaard P. Risk of newly diagnosed type 2 diabetes is reduced in users of alendronate. Calcif Tissue Int 2011; 89: 265-70.

[59]   Foresta C, Strapazzon G, De Toni L, *et al.,* Evidence for osteocalcin production by adipose tissue and its role in human. J Clin Endocrinol Metab 2010; 95: 3502-6.

[60]   Schafer AL, Sellmeyer DE, Schwartz AV, *et al.,* Changes in undercarboxylated osteocalcin is associated with changes in body weight, fat mass, and adiponectin: parathyroid hormone (1-84) or alendronate therapy in postmenopausal women with osteoporosis (the PaTH study). J Clin Endocrinol Metab 2011; 96: E1982-9.

CHAPTER 5

Emerging Therapeutic Agents for the Treatment of Type 2 Diabetes Mellitus

Shivaprasad Channabasappa[1,*] and Krishnan M. Prasanna Kumar[2]

[1]Department of Endocrinology and Metabolism, Vydehi Institute of Medical Sciences & Research Centre, Banglore, India and [2]Center for Diabetes and Endocrine Care, Banglore, India

Abstract: The worldwide epidemic of type 2 diabetes (T2D) and the limitations of currently available agents for its treatment have fueled the quest for new therapeutic agents and targets for the treatment of this incurable disease. Despite the availability of various pharmacologic agents, the treatment of patients with type 2 diabetes mellitus remains suboptimal and currently available therapies do not significantly improve β-cell function. Available therapies for diabetes are burdened by side-effects, and they do not adequately address the multiple defects in glucose homeostasis thought to play key roles in the pathophysiology of T2D. Most of the current therapies for T2D were developed in the absence of defined molecular targets. Intensive molecular research and the better understanding of its pathophysiology of T2D have led to the expansion of pharmacologic repertoire to target novel physiologic mechanisms. These drugs differ from established therapies in their mechanisms of action and molecular targets and hence complement them. Incretin-based therapies are a prime example of such novel agents and constitute a priced addition to the pharmacologic inventory of T2D. SGLT2 inhibitors, the latest addition to the T2D treatment has a unique glucose lowering mechanism and also contributes to weight loss. Emerging new molecules, which hold promise for the treatment of T2D include glucokinase activators, dual PPAR α/γ agonists, G protein-coupled receptor 119 agonists, interleukin-1 β inhibitors, protein tyrosine phosphatase 1B inhibitors, gluconeogenesis inhibitors, glycogen synthase kinase-3 inhibitors and 11 β-hydroxysteroid dehydrogenase type 1 inhibitors. QNEXA and lorcaserin are experimental molecules being tried to induce weight loss and also have the potential to reduce blood glucose levels. Stem cell therapies are being explored as a new potential treatment for T2D. Bariatric and metabolic surgeries are also being pursued as therapeutic options for T2D. Expectations from these new therapies in the pipeline for diabetes are very high, and they may fill the gap in the unmet needs of diabetes management. These emerging innovative therapeutic modalities are the focus of discussion in this review.

Keywords: Type 2 diabetes, emerging therapies, unmet needs, incretins, SGLT2 inhibitors, molecular targets, obesity, β-cell function, insulin resistance.

*Address correspondence to Shivaprasad Channabasappa: Department of Endocrinology & Metabolism, Vydehi Institute of Medical Sciences & Research Centre, Bangalore, India; Tel: 08028413381; E-mail: shvprsd.c@gmail.com

Atta-ur-Rahman (Ed)

INTRODUCTION

Type 2 diabetes (T2D) results from a complex interaction between genetic, environmental and lifestyle factors and is characterized by insulin resistance, pancreatic β-cell dysfunction and reduced β-cell mass. Overweight, visceral obesity, genetic factors, sedentary lifestyle and dietary habits are major contributors to the development of insulin resistance. Insulin resistance is present for many years before the onset of T2D, and if it not addressed properly leads to impaired glucose tolerance. Failure of β-cells to compensate for insulin resistance by sufficient insulin secretion leads to progression of impaired glucose tolerance to overt type 2 diabetes. Insulin resistance is characterized by hyperinsulinemia and is also associated with atherosclerosis, hypertension, abnormal lipid profile and adverse cardiovascular risk factors. Traditional triumvirate concept of pathophysiology of T2D entails three core pathophysiologic defects in its development - insulin resistance in muscle and liver and β-cell failure. However, this triumvirate concept is simplistic and fails to recognize other core defects involved in pathophysiology of T2D. It is now widely acknowledged that β-cell failure occurs much earlier during T2D and was underestimated previously.

Abnormalities in incretin hormones with reduced incretin effect, hyperglucagonemia, raised concentrations of other counter-regulatory hormones, alterations in the fat cell and adipokines also contribute to development of T2D. Chronic hyperglycemia (glucotoxicity), non-esterified fatty acids (lipotoxicity), oxidative stress, inflammation and amyloid formation all contribute to the decline in β-cell function. Pancreatic α-cell dysfunction results in non-suppressed glucagon secretion in the post prandial period with consequent hyperglycemia. CNS involvement in T2D pathophysiology has been implicated with increased appetite due to defective hypothalamic appetite regulation, alterations in melatonin and circadian genes (contributing to reduce insulin secretion) and cerebral insulin resistance. These changes could result in increased hepatic glucose output and reduced muscle glucose uptake. Dr. Ralph DeFranzo of the University of Texas coined the term 'Ominous Octet' to encapsulate additional metabolic defects in the development of T2D. According to the ominous octet concept for pathophysiology of T2D, in addition to the muscle, liver, and beta cell (triumvirate), there are five more factors, which play important roles. These five factors are fat cells (increased lipolysis), gastrointestinal tract (incretin deficiency/resistance), α-cells (increased glucagon levels), kidneys (increase in glucose reabsorption), and the brain (increased appetite and possible contribution to insulin resistance).

The most important implication of the ominous octet is that no single drug can adequately address all these eight defects, and hence combination pharmacotherapy will be required to address the multiple pathophysiological defects. Existing pharmacotherapy for T2D does not counter progressive β-cell dysfunction, and as a consequence their benefits are not sustained in long term. Most of the current therapies for T2D were developed in the absence of defined molecular targets and are burdened by side effects. Intensive molecular research and the better understanding of its pathophysiology of T2D have led to the discovery of novel molecular targets. Given the limitations of current pharmacologic repertoire, fresh treatments that will address β-cell failure, sustain glycemic control, improve insulin action without causing hypoglycemia and weight gain are required. These novel drugs should also have a favorable effect on cardiovascular outcomes.

In this review, we focus on the emerging therapies that have novel mechanism of actions and are being developed for patients with type 2 diabetes. Newer developments in incretin-based therapies are covered first followed by SGLT2 inhibitors. Emerging new molecules, which hold promise for the treatment of T2D, including G protein-coupled receptor 119 agonists, glucokinase activators, interleukin-1 β inhibitors, protein tyrosine phosphatase 1B inhibitors, gluconeogenesis inhibitors, glycogen synthase kinase-3 inhibitors and 11β-hydroxysteroid dehydrogenase type 1 inhibitors are covered in subsequent sections. A bird's eye view is provided on dual PPAR α/γ agonists. Bariatric and metabolic surgeries and anti-obesity drugs QNEXA and lorcaserin which have the potential to reduce blood glucose levels are also reviewed. Finally, stem cell therapies being explored as a new potential treatment for T2D is covered.

EMERGING INCRETIN BASED THERAPIES

The augmentation of insulin secretion following an oral glucose load when compared to an isoglycemic intravenous glucose load is referred to as the incretin effect. This incretin effect is due to gut-derived factors called incretins, which are released in response to oral nutrient ingestion. The incretin effect accounts for 50–70 % of insulin secreted after meals and in humans, two hormones—gastric inhibitory polypeptide (GIP) and glucagon- like peptide-1 (GLP-1)—account for the incretin effect. GIP and GLP-1 are polypeptides that are released from the gut in response to food ingestion and potentiate insulin secretion in a glucose dependent manner. GIP is secreted from K-cells in the duodenum, while GLP-1 is secreted from L-cells primarily located in the distal ileum. GLP-1 has additional effects, including reducing glucagon secretion, slowing gastric emptying, inducing satiety, causing weight loss, as well as trophic effects on the pancreas.

Incretins are rapidly inactivated by dipeptidyl peptidase 4 (DPP-4), which cleaves the active peptide at the alanine residue that is penultimate to the N terminus. DPP-4 is widely expressed, especially by endothelial cells lining vessels that drain from the intestinal mucosa. This is responsible for rapid inactivation and short circulating half-life of incretins (<2 min for GLP-1 and 5–7 min for GIP). To extend the half-life, DPP-4-resistant GLP-1 analogues with GLP-1-receptor (GLP-1R) agonist properties have been developed (exenatide, liraglutide). Another strategy has been to increase endogenous GLP-1 by highly specific DPP-4 inhibitors (sitagliptin, vildagliptin, saxagliptin). Two GLP-1 analogues, exenatide and liraglutide are approved for use in type 2 diabetes. Four DPP-4 inhibitors are currently approved for the management of type 2 diabetes—sitagliptin, vildagliptin, saxagliptin and linagliptin. Due to its short half-life, exenatide must be administered twice-daily by subcutaneous (SC) injections.

BYDUREON, sustained release formulation of exenatide was approved by FDA in January 2012. Phase III clinical trials of some short acting (lixisenatide) and sustained-release drugs (taspoglutide, albiglutide, and CJC-1134-PC) are in progress.

Sustained Release Formulations of Incretin Mimetics -Incretin Mimetics Fused to Carriers

Sustained release formulations of incretin mimetics have been achieved by fusing them to polymeric microspheres, albumin or Fc Fragment. Sustained release exenatide formulation (BYDUREON) has been achieved by attaching biodegradable polymeric microspheres of poly (D, L-lactic-co-glycolic acid) to exenatide [1]. Albiglutide is another longer acting incretin mimetic which consists of two copies of a 30 amino acid sequence (7–36) of human GLP-1 that is made DPP-IV resistant by alanine to glycine conversion at amino acid 8. It is subsequently fused with human albumin and provides a long half-life of ~6 – 8 days [2, 3]. Taspoglutide (R1583/BIM51077) is 8-(2-methylalanine)-35-(2-methylalanine)-36-L-argininamide derivative of the amino acid sequence 7–36 of human GLP-I. It is another long-acting GLP-1 analogue which can be administered once weekly [4]. Dulaglutide (LY-2189265) is a novel, long-acting glucagon-like peptide 1 (GLP-1) derivative, which consists of GLP-1(7-37) covalently linked to a modified Fc fragment of human IgG4. Fusion of GLP-1 to this larger 'carrier' moiety protects the GLP-1 moiety from inactivation by DPP 4 and slows its *in vivo* clearance [5, 6]. Dulaglutide has a t1/2 in humans of up to 90 h and a flat profile with no burst effect like taspoglutide making it an ideal candidate for once-weekly dosing. CJC-1134-PC, is a long-acting Exendin-4 analogue consisting of modified Exendin-4 analogue conjugated to recombinant human albumin. It is based on ConjuChem's PC-DAC™ technology. Its half-life of about eight days is similar to that of circulating albumin. The most desirable feature of the sustained-release formulations is the advantage of intermittent administration of once weekly or longer. The durable DPP-4 resistance of these GLP-1 analogues has been achieved by use of different methods of preparation.

Exenatide- LAR (BYDUREON)

Twice daily injections are one of the major drawbacks of exenatide. This drawback has been successfully overcome by formulating Exenatide- LAR (BYDUREON), based on Medisorb technology. It is administered by subcutaneous injection once weekly at a dose of 2 mg and therapeutic concentrations are achieved in two weeks, and steady-state concentrations are reached between 6-10 weeks.

DURATION (Diabetes therapy Utilization: Researching changes in HbA1c, weight and other factors Through Intervention with exenatide once weekly), is a Phase III clinical program to evaluate exenatide-LAR. DURATION-1 was a 30-

week non-inferiority study that compared exenatide LAR 2 mg given as a once-weekly dose with exenatide 10 μg given twice daily. Patients given exenatide LAR weekly had a change in HbA1c of -1.9 % compared with -1.5 % in patients given twice-daily injections and weight reduction was similar between the two groups (-3.7 kg and -3.6 kg). Reduction in FPG was more with the once-weekly treatment arm (-2.3 mmol/L *vs.* -1.4 mmol/L), whereas reduction in PPG was greater in patients treated with exenatide twice daily (-5.3 mmol/L *vs.* -6.9 mmol/L). Nausea or vomiting was less common in the LAR, although more LAR-treated patients reported injection-site pruritus [7].

Based on published data comparing exenatide LAR with twice daily injections, LAR appears to cause greater HbA1c reduction with improved treatment satisfaction. After thirty weeks of treatment exenatide LAR lowers HbA1c by -1.5 % to -2 %, and effect on HbA1c appears to be sustained after two years of treatment. By comparison, twice-daily exenatide was shown in clinical trials to lower HbA1c by about -1 %. Gastro-intestinal adverse effects are lower with once weekly preparations [8]. In the pooled data from the DURATION-1, -2 and -3 trials, overall incidence rates of adverse events (AEs) (77 % *vs.* 71 %); serious AEs (4 % *vs.* 5 %) and discontinuations due to serious AEs (0.7 % *vs.* 0.5 %) were similar for BYDUREON *versus* pooled comparators- sitagliptin, pioglitazone and insulin glargine. The BYDUREON had lower incidence of hypoglycemic events and the incidence of pancreatitis, renal dysfunction/dehydration and thyroid-neoplasms were similar for BYDUREON when compared with pooled comparators (Amylin and Eli Lilly press release, presented at ADA 2010). Long-term safety and efficacy data on BYDUREON have not yet been published.

Albiglutide (Syncria)

Albiglutide 30 mg weekly was selected as the initial dose for the phase III programme, named HARMONY Clinical Research Program. In Harmony 7, a head-to-head study comparing albiglutide to liraglutide, HbA1c reduction was -0.78 % for patients receiving albiglutide compared to -0.99 % with liraglutide. Nausea and vomiting rates were lower in patients receiving albiglutide compared to those receiving liraglutide. However, weight loss in patients receiving albiglutide (-0.62 kg) was lower than that observed with liraglutide (-2.21 kg). In

Harmony 6, a study involving diabetic patients on insulin glargine, albiglutide produced clinically significant reductions in HbA1c (-0.82 %) from baseline compared to pre-prandial lispro insulin (-0.66 %) after 26 weeks of treatment [2, 3, 9-11]. Results of other Phase III trials on albiglutide are awaited.

Taspoglutide

Taspoglutide was evaluated in a 24-week multicenter Phase III trial involving 373 drug naive patients with T2D. Weekly subcutaneous taspoglutide 10 or 20 mg was compared with placebo. Taspoglutide 20mg resulted in HbA1c reduction from baseline of -1.18 % and weight loss of -2.25kgs. Gastrointestinal adverse events and injection site reactions were more common with taspoglutide than placebo [12]. Taspoglutide phase III trials were stopped in September 2010 by Roche because of higher than expected discontinuation rates of taspoglutide treated patients, mainly due to GI tolerability and severe hypersensitivity reactions.

Dulaglutide

In Phase II clinical trials dulaglutide, demonstrated a dose-dependent reduction in HbA1c of up to 1.5 % and weight reduction of up to 2.5kg compared with placebo. GI disturbances, including nausea, diarrhea and abdominal distension were the most frequently reported adverse events. Phase III clinical trials are examining the efficacy and adverse effects of dulaglutide [5, 6, 10-11].

CJC-1134-PC

Data from preclinical and phase I/II studies demonstrated that CJC-1134-PC retains the full spectrum of GLP-1 receptor-dependent actions and also has a positive tolerability profile [13]. In mice, CJC-1134-PC caused a dose-dependent reduction in the glycemic excursions and increased plasma insulin/glucose ratios in response to either glucose challenge. Once-daily subcutaneous injections of 100 nmol/kg of CJC-1134-PC for eight days delayed diabetes progression in db/db mice [14].

In Phase II clinical studies in humans starting doses of 1.5 mg weekly and 1.5 mg twice weekly (3 mg weekly) subcutaneous injections were used. In these trials when it was administered once weekly it showed a dose-dependent reduction in HbA1c of up to -1.4 %. The reduction in weight in the CJC-1134-PC groups was modest. Its unpretentious efficacy in weight reduction may be due to its inability to cross the blood–brain barrier due to conjugation to albumin and hence not having access to CNS regions regulating appetite and body weight. GI disturbances, including nausea and vomiting were the most frequently reported adverse events [15 -17]. Other Phase II clinical trials are currently underway.

Comparison between Long Acting and Short Acting Incretin Therapies

In general, a greater reduction in hemoglobin A1c (HbA1c) and fasting plasma glucose has been found with the once-weekly GLP-1 receptor agonists compared with exenatide BID, while the effect on postprandial hyperglycemia was inferior to that of exenatide. Once-weekly GLP-1 receptor agonists are associated with a larger increase in fasting insulin, larger decrease in fasting glucagon, less nausea (except taspoglutide) and lower hypoglycemia. Weight loss is lower when compared to exenatide. Overall, higher treatment satisfaction for patients has been noted with long acting agonists because of their ease of use, less nausea and need for less-frequent dosing requirement. However, they are associated with a larger increase in heart rate and reduced effects on gastric emptying [10, 11, 17-19].

Lixisenatide – A New Short Acting Exendin–4 Analogue

Lixisenatide (also known as AVE0010 and ZP10) is a synthetic modified version of exendin–4 which is resistant to degradation by DPP-4 as a result of modification at C-terminal region with six new lysine residues and deletion of one proline. The half-life of lixisenatide is only 2–4 hours, but its strong binding affinity to the GLP-1 receptor allows once-daily dosing. In Wistar rats, lixisenatide significantly improved glucose-stimulated insulin secretion and its effect on insulin secretion was greater than that observed with GLP-1, consistent with its greater affinity to the GLP-1 receptor [20, 21].

In a Phase II study, two dose increment regimens of lixisenatide 20 μg given once daily significantly improved the glycemic control in mildly hyperglycemic patients with T2D on metformin. Both FPG and post prandial glucose values were significantly reduced when compared to placebo. Lixisenatide had a pronounced postprandial effect and reduced glycemic excursion by 75 % [22]. In Phase II trials, lixisenatide 20 μg once daily restored first-phase insulin release and also improved second-phase the insulin response. This dose reduced HbA1C by 0.75 % from a baseline value of 7.55 %. The Phase III clinical trial program for lixisenatide known as the GetGoal program began in 2008. In GetGoal-Mono, a Phase III study involving 361 T2D patients not on glucose-lowering therapy given lixisenatide or placebo, once-daily lixisenatide significantly improved HbA1c (least squares mean change *vs.* placebo of up to 0.66 %) and significantly more lixisenatide patients achieved HbA1c <7.0 % compared to placebo. It had a pronounced postprandial glucose lowering effect with 75 % reduction in the glucose excursions. Mean decrease in body weight was ~2 kgs, and the most common adverse events were gastrointestinal [23, 24]. Results of other Phase III studies are pending.

Dual-Acting Peptides Based on Incretins and Glucagon

Glucagon-like peptide-1 (GLP-1) and glucagon share a common precursor, pre-pro glucagon, which is processed in a tissue-specific manner leading to production of glucagon in the pancreas and GLP-1 in the intestine. GLP-1 and glucagon have contrasting effects on plasma glucose. Plasma glucose is elevated by glucagon as a result of increased glycogenolysis and gluconeogenesis in the liver. GLP-1 induces glucose-dependent insulin secretion in the pancreas while suppressing glucagon secretion. Combination of GLP-1 receptor agonists and glucagon receptor antagonists can lead to incremental reductions in plasma glucose and can also have other complementary effects in glycemic control. Dual-acting Peptides based on GLP-1 agonism and glucagon antagonism arouse interest as potential new therapeutic targets for the treatment of T2D. Pegylated Dual-acting Peptide for Diabetes is one such molecule under investigation for the treatment of T2D.

Pegylated Dual-acting Peptide for Diabetes (PEG-DAPD)

Dual-acting peptide for diabetes (DAPD) is a GLP-1/glucagon hybrid peptide that exhibits both GLP-1 receptor agonist activities and glucagon receptor antagonistic activities. It activates the GLP-1 receptor while inhibiting the glucagon receptor. To prolong the duration of action, DAPD was conjugated with a single polyethylene glycol (PEG) at C-terminal cysteine, which is called PEG-DAPD. In mice, PEG-DAPD reduced area under the glucose curve during glucose tolerance testing, and glucose-lowering effects of PEG-DAPD were more prolonged (up to at least 17 hours) relative to DAPD. In diabetic db/db mice, PEG-DAPD showed extended metabolic stability and significantly reduced fasting blood glucose levels compared to controls [25, 26]. Although no human studies have been published PEG-DAPD has the potential to be developed as a novel incretin therapy to treat T2D, with augmented efficacy and prolonged action.

ZP2929

ZP2929 is another dual acting peptide like PEG-DAPD, but unlike the latter it acts as an agonist on both GLP-1 and glucagon receptors. It is based on a naturally occurring gut peptide hormone called oxyntomodulin. Oxyntomodulin levels are high postprandially. The small intestine is the site of production of oxyntomodulin and exerts its biological effects by activating both the GLP-1 receptor and the glucagon receptor. Its effects in humans include improved glucose tolerance and weight loss. Glucagon, although counteracts GLP-1's effect on blood glucose induces fat breakdown and by doing so it is postulated to produce greater weight loss than GLP-1.

Acting as an agonist on both the GLP-1 and glucagon receptors, ZP2929 has been shown in preclinical studies the ability to achieve glycemic control and cause weight loss. Once-daily subcutaneous administration of ZP2929 is under development for improvement of glycemic control and induction of weight loss in obese type 2 diabetes patients. In db/db mice, twice daily SC injections of ZP2929 for 21 days reduced blood glucose levels without any weight gain [27]. Similarly, ZP2929 improved glycemic control, reduced body weight markedly and improved lipid profile in murine models of obesity and insulin resistance [28]. In June 2011, Zealand Pharma and Boehringer Ingelheim entered an exclusive global license and collaboration agreement for the development of ZP2929 and other DAPDs for treatment of diabetes.

Newer DPP4 Inhibitors

Alogliptin

Alogliptin benzoate is a potent and selective quinazolinone based noncovalent inhibitor of DPP4 administered orally. Its selectivity for DPP-4 is >10,000-fold greater than that of the other DPP isozymes 2/8/9. Plasma DPP-4 activity was inhibited by > 80 % by Alogliptin after 24 hours, and hence it is administered once daily. It is primarily excreted unchanged in the urine [29].

Alogliptin has been studied in phase II and III clinical trials mainly as an add-on drug to other OADs. In type 2 diabetic patients inadequately controlled by metformin, alogliptin added to pioglitazone produced additional HbA1C reduction of 0.5 % when compared to pioglitazone alone [30]. Alogliptin in doses of 12.5 mg and 25 mg when used as an add-on to metformin therapy for 26 weeks caused HbA1c reduction of 0.6 % by both doses compared to the placebo [31]. Combination therapy of alogliptin added to glyburide caused significant HbA1c reductions across increasing doses of alogliptin: –0.39 % with 12.5 mg, and –0.53 % with 25 mg dose [32]. 26-wk treatment with alogliptin combined with pioglitazone, in patients not controlled on metformin produced additive effects on reduction in FPG and HBA1C [33]. In newly diagnosed drug naïve Japanese patients, alogliptin was effective and non-inferior to traditional Japanese diet as an initial therapeutic option [34]. As an add-on therapy in Japanese T2D patients uncontrolled on voglibose plus diet and exercise, alogliptin 25 mg reduced HbA1c by 0.93 % compared to 0.06 % increase with placebo [35]. EXAMINE- an ongoing trial will define the CV safety profile of Alogliptin in T2DM patients at high risk for CV events [36]. Alogliptin secured FDA approval in Jan 2013.

Gemigliptin (LC15-0444)

Gemigliptin (LC15-0444) is a new dipeptidyl peptidase-4 (DPP-4) inhibitor being co-developed by LG Life Sciences and Double-Crane Pharmaceutical Co. It is a selective and competitive inhibitor of DPP IV. Studies on its pharmacokinetics and pharmacodynamics have revealed that it permits once-daily dosing regimen [37]. Gemigliptin is a competitive, reversible inhibitor of DPP-4 (IC50 = 16 nM) with excellent selectivity for DPP-4 over other relevant human proteases. Gemigliptin is rapidly absorbed after single oral dosing and has an elimination half-life of 3.6 hours in the rats. In a Phase II study, LC15-0444 monotherapy (50 mg for 12 weeks) reduced the HbA1c by 0.98 % as compared to placebo in Korean subjects with type 2 diabetes. It also improved FPG level, OGTT results, β-cell function and insulin sensitivity measures in them [38]. Ongoing clinical trials are evaluating the safety and efficacy of gemigliptin in combination with metformin and other agents in T2D.

SODIUM GLUCOSE COTRANSPORTER – 2 (SGLT2) INHIBITORS

In the kidney, the majority of the filtered glucose is reabsorbed by the sodium glucose co transporter – 2 (SGLT2). SGLT proteins are encoded by the solute carrier five (SLC5) subfamily of sodium/substrate symporter genes. SGLT1 and SGLT2 are the most well characterized products of SCL5 genes. In kidney, SGLT1 is expressed in the distal segments (S2–3) of the proximal convoluted tubule, and SGLT2 is almost exclusively expressed in the first segment (S1) of the proximal tubules in the kidney (Table **1**). SGLT1 is a high-affinity, low-capacity transporter and accounts for only about 10 % of renal glucose reabsorption and SGLT2 is a low-affinity, high-capacity transporter and accounts for more than 90 % of renal glucose reabsorption. It has been found that patients with T2D express a higher number of SGLT2 than healthy individuals, and renal glucose reabsorption is elevated in patients with T2D [39, 40]. SGLT2 inhibitors represent a new mechanism to reduce hyperglycemia, and their action doesn't affect insulin secretion or its action. They lower fasting and postprandial glucose levels, improve glycemic control, and cause weight loss with a low risk of hypoglycemia. They reduce plasma glucose values irrespective of patients' glycemic status and act independently of insulin. Their osmotic diuretic effect can be beneficial in patients with congestive heart failure. They may reduce BG values even in type 1 diabetes.

Table 1: Comparison of SGLT-1 and SGLT-2 Transporters

	SGLT-1	SGLT-2
Major expression site	Small intestine	Kidney
Expression within kidney	Late proximal straight tubule (S3 Segment)	Early proximal convoluted tubule (S1 segment)
Glucose transport capacity	Low	High
Affinity for glucose	High ($Km = 0.4$ mM)	Low ($Km = 2$mM)
Contribution to renal glucose reabsorption	~10 %	~90 %

Km- Michaelis constant.

It was known more than 50 years ago that a naturally occurring phenol glycoside called phlorizin isolated from the bark of apple trees could increase urinary glucose excretion by inhibiting glucose reabsorption. In the 1990s it was shown that phlorizin was a potent, competitive inhibitor of both SGLT1 and SGLT2. Lack of selectivity towards SGLT2 and toxicity precluded phlorizin as a drug candidate for T2DM. Phlorizin binds to SGLT2 transporters *via* the glucoside moiety and the 'O'-linked phenolic distal ring is responsible for its SGLT2 inhibitory properties. However, O-linkage is a metabolic target for β-glucosidase enzymes, which inactivate it and this limitation has been overcome by the candidate SGLT2 inhibitors that employ a C-glucoside linkage. Several SGLT2 inhibitors based on the glucoside structure of phlorizin have since been discovered, and dapagliflozin is the best characterized inhibitor among them [39, 40].

SGLT2 inhibitors can be divided into three main categories- C glucosides, O glucosides and antisense oligonucleotides to inhibit expression of SGLT2. C glucosides include dapagliflozin, canagliflozin, ASP-1941, BI-10773, LX-4211 and DSP-3235. O-glucosides are sergliflozin, remogliflozin, AVE-2268 and YM-543. ISIS-388626 is an antisense oligonucleotide.

Dapagliflozin

Dapagliflozin was identified by Bristol–Myers Squibb as a potent and selective SGLT2 inhibitor using structure–activity relationship (SAR) studies. It is a C-aryl glucoside with 1200-fold selectivity for SGLT2 over SGLT1and 30-times greater potency against SGLT2 in humans when compared to phlorizin. C-aryl glucoside linkage confers resistance to β-glucosidase enzymes. It has a half-life of 12.5 hours and is rapidly absorbed following oral administration and reaches the max concentration in two hours. Renal glucose excretion increases in a dose-dependent fashion and reaches a plateau at the 20 mg/day dose. The proposed dose is 10 mg, taken once daily at any time of day [39-41].

Dapagliflozin has been tested extensively in phase II and III studies. Initial studies used up to 100 mg/d, but subsequent studies have used three doses of 2.5 mg, 5 mg and 10 mg/day. Komoroski *et al.,* evaluated the efficacy of dapagliflozin (three doses 5-, 25-, or 100-mg PO) in comparison to placebo in a 14-day study randomized trial in patients with T2D. 100 mg/d reduced fasting glucose by 9.3 % on day 2 and dose-dependent reductions in FPG were observed on day 13 with the 5 mg (-11.7 %), 25 mg (-13.3 %), and 100 mg (-21.8 %) doses as compared with placebo. All 3 doses produced significant improvements in oral glucose tolerance test (OGTT) improved as compared with placebo. On day 14, dapagliflozin 100mg/d caused glycosuria of nearly 70 g/d [42]. Two 24-weeks randomized, double-blind trials compared dapagliflozin plus metformin, dapagliflozin alone and metformin alone in treatment-naïve T2D patients. Dosage of dapagliflozin was 5 mg in Study 1 and 10 mg/d in Study 2. HbA1c reduction in combination arms, dapagliflozin and metformin arms in the two studies were 2 % and 1.9 %, 1.2 % and 1.45 %, 1.3 % and 1.45 % respectively. Combination therapy was statistically superior to monotherapy in FPG reduction and dapagliflozin 10 mg/d was non inferior to metformin in study 2. Urinary tract and genital infections were more common with dapagliflozin [43].

Published Phase III trials of dapagliflozin are summarized in Table **2** [44-49]. A 48 week long-term efficacy study compared dapagliflozin (three doses 2.5, 5, 10 mg/d) *vs.* placebo in 808 patients with T2D uncontrolled on high doses of insulin (at least 30 U/d). After 24 weeks, mean HbA1c decreased by 0.79 % to 0.96 % with dapagliflozin 5 and 10mg doses compared to 0.39 % with placebo. The daily dose of insulin decreased by 1.95 units with dapagliflozin 10mg/d, and increased by 5.65 units with placebo. The body weight decreased by 0.92-1.61 kgs with dapagliflozin, and increased by 0.43 kg with placebo. Compared with the placebo group, patients in the pooled dapagliflozin groups had a higher rate of hypoglycemic episodes (56.6 % *vs.* 51.8 %), genital infection (9.0 % *vs.* 2.5 %) and urinary tract infection (9.7 % *vs.* 5.1 %) [46].

A 52-week noninferiority trial compared dapagliflozin with glipizide as an add-on to patients receiving metformin monotherapy T2D (baseline mean HbA1c 7.7 %). Mean HbA1c reduction with dapagliflozin (-0.52 %) was similar to that with glipizide (-0.52 %). Dapagliflozin caused weight loss and produced less hypoglycemia than glipizide [48]. Patients with type 2 diabetes inadequately controlled on pioglitazone, were randomized to receive dapagliflozin 5mg (n = 141) or 10 mg (n = 140) or placebo (n = 139). Dapagliflozin 5 and 10 mg groups reduced by -0.82 and -0.97 % compared to -0.42 % for placebo. Patients receiving pioglitazone alone had a greater weight gain (three kgs) than those receiving dapagliflozin plus pioglitazone (0.7-1.4 kg) and at week 48 more edema than those receiving dapagliflozin plus pioglitazone. Hypoglycemia was rare in both groups [49].

Table 2: Phase III Clinical trials of dapagliflozin

Subjects	Duration and Subjects	Dapagliflozin Dose	Comparator	Fpg (Mg/Dl)#	HBA1C#	Uti/genital infections#
Treatment-naïve diabetic patients [44]	24 weeks n=485	2.5, 5 and 10 mg/d	Placebo	↓15–29 ↓4.9	↓0.58–0.89 % ↓0.23 %	4.6 -12.5 %* 4 %*
Patients uncontrolled on metformin ≥1.5g/d [45]	24 weeks n=534	2.5, 5 and 10 mg/d	Placebo	↓15–29 ↓4.1	↓0.67–0.84 % ↓0.3 %	8 -13 %* 5 %*
Patients uncontrolled on insulin± OAD [46]	48 weeks n=808	2.5, 5 and 10 mg/d	Placebo	↓12–17 -	↓0.79–1.01 % ↓0.47 %	5.4 -7.5 %* 4.1 %*
Patients uncontrolled on Glimeperide 4mg/d [47]	24 weeks n=597	2.5, 5 and 10 mg/d	Placebo	↓17–28 ↓4.9	↓0.58–0.82 % ↓0.13 %	3.9 -6.9 %* 6.2 %*
Patients uncontrolled on metformin ≥1.5g/d [48]	52 weeks n=814	2.5, 5 and 10 mg/d	Gipizide-5, 10&20mg/d	↓22.34 ↓18.72	↓0.52 % ↓0.52 %	10.8 %* 6.4 %*
Treatment-naïve/patients on OAD shifted to pioglitazone monotherapy [49]	48 weeks n=420	5 and 10 mg/d	Placebo	↓23–33 ↓ 13	↓0.95–1.21 % ↓0.54 %	4.6 -12.5 %^ 4 % ^

#Dapagliflozin *vs.* comparator, *Urinary tract infections, ^ genital infections.

FDA rejected the new drug application of dapagliflozin in 2011 because of evidence of possible liver damage and the potential link with breast and bladder cancer. Data regarding long-term safety, including urinary tract/genital infections and cardiovascular safety are needed before it finds a place in the algorithm of T2D management.

Canagliflozin

Canagliflozin, a thiophene derivative is a c–glucoside SGLT-2 inhibitor in phase III trials in the treatment of T2D. It is a potent and selective SGLT2 inhibitor. It has 200-fold selectivity for SGLT2 over SGLT1. Canagliflozin pharmacokinetics is dose-dependent, and the elimination half-life ranges from 12 -15 h. In obese animal models, canagliflozin increased urinary glucose excretion and decreased plasma glucose, renal threshold for glucose excretion, body weight, epididymal fat and liver weight [50]. In a 28-day trial of twenty nine patients inadequately controlled on insulin +OAD, canagliflozin {100 mg QD or 300 mg twice daily (BID)} was compared to placebo. HbA1c reduction was 0.19 % with placebo, 0.73 % with 100 mg QD, and 0.92 % with 300 mg BID of canagliflozin. Both doses of canagliflozin caused weight loss as compared to placebo [51].

A 12 week dose-ranging study in 451 T2DM subjects compared the efficacy of canagliflozin 50, 100, 200, or 300 mg once daily (QD) or 300 mg twice daily (BID) with, sitagliptin 100 mg QD, or placebo in patients inadequately controlled with metformin monotherapy. HbA1c reductions were −0.79, −0.76, −0.70, −0.92, and −0.95 % for canagliflozin 50, 100, 200, 300 mg QD and 300 mg BID, respectively compared to −0.22 % for placebo (all P < 0.001) and −0.74 % for sitagliptin. Canagliflozin reduced FPG by −16 to −27 mg/dL, and body weight by −2.3 to −3.4 %. There was an increase in symptomatic genital infections with canagliflozin (3–8 %) *versus* placebo and sitagliptin (2 %) [52]. Canagliflozin was approved by FDA in March 2013.

Sergliflozin etabonate and Remogliflozin etabonate are O glucoside derived selective SGLT-2 inhibitors. They did not undergo further development after phase II stage. Ipragliflozin (ASP1941) is a novel C-glucoside with benzothiopene structure, which is a potent and selective inhibitor of SGLT-2. *In vitro*, the potency of ipragliflozin to inhibit SGLT2 and SGLT1 and stability were assessed. In experimental studies, Ipragliflozin selectively inhibited human, rat, and mouse SGLT2 at nanomolar ranges and caused a dose-depend increase in urinary glucose excretion for over 12 hours following single oral dose [53]. In humans, Ipragliflozin was rapidly absorbed following oral administration and reached the peak concentration in 1.3 hours after the last dose with a mean elimination half-life of 12 hours. Ipragliflozin increased UGE in a dose-dependent manner up to a maximum of approximately 59 g of glucose excreted over 24 hours [54]. It is being investigated in phase III clinical trials. Empagliflozin (BI-10773) is a potent and competitive SGLT-2 inhibitor with an excellent selectivity profile and display's high selectivity window for SGLT-2 inhibition over SGLT-It has shown encouraging results in phase II trials and phase III studies are currently underway [55].

ISIS 388626 – A SGLT2 Antisense Drug

ISIS 388626 is a very potent and selective antisense inhibitor of SGLT2 that reduces renal SGLT2 expression by >80 % in multiple species, including mice, rats and monkeys. It targets the mRNA of SGLT-2 gene reducing the production SGLT-2 protein. Another unique feature of ISIS 388626 is its short 12 nucleotide length, which confers it an unusually high potency. It specifically inhibits the production of SGLT2 in the kidney tissue, without any effect on SGLT1 production. ISIS 388626 is in early stage of clinical development for type 2 DM [40].

LX4211

LX4211 is a dual inhibitor of sodium-glucose co transporters 1 and 2 under development for T2D treatment. In early clinical studies it demonstrated beneficial effects on postprandial glucose levels, triglyceride levels and other parameters of glycemic control in healthy subjects. A single dose of LX4211 significantly increased circulating levels of GLP-1 (active and total) and PYY, important regulators of glycemic and appetite control. A single dose mechanistic and pharmacokinetic study in combination with sitagliptin (Januvia®) demonstrated complementary effects, increasing active GLP-1 and lowering post prandial glucose.

G PROTEIN-COUPLED RECEPTOR 119 (GPR119) AGONISTS

G protein-coupled receptor 119 (GPR119) is predominantly expressed in the β-cells of pancreatic islets and L-cells of the gut, and is involved in insulin and incretin hormone release. It is encoded by the GPR119 gene in humans and plays an important role in regulation of incretin and insulin hormone secretion. Activation of GPR119 leads to increase in intracellular cAMP *via* adenylate cyclase activation, which results in insulin release from the β-cells and GLP-1 release from the L-cells. Both these effects lead to augmentation of glucose stimulated insulin secretion (GSIS). GPR119 agonists thus mediate a unique nutrient-dependent dual elevation of both insulin and GLP-1 levels *in vivo*. Activation of the receptor has also been shown to reduce food intake and body weight gain in rats. In addition, there are some data to suggest that they may have a role in β- cell preservation also [56, 57]. GPR119 agonists either as a mono therapy or in combination with DPP-IV inhibitors are promising agents for T2D (Fig. **1**). Quite a few small molecule oral GPR119 agonists have been developed, but most lack the ability to adequately directly preserve the β-cell function limiting their clinical utility. A novel structural class of small-molecule GPR119 agonists consisting of 2, 4, 6-tri-substituted pyrimidine cores with promising potential for the treatment of T2D have been developed. AS1269574 is the first compound to be evaluated in this group. It is capable of inducing GSIS *in vitro* and *in vivo* and improved glucose tolerance in normal mice [58].

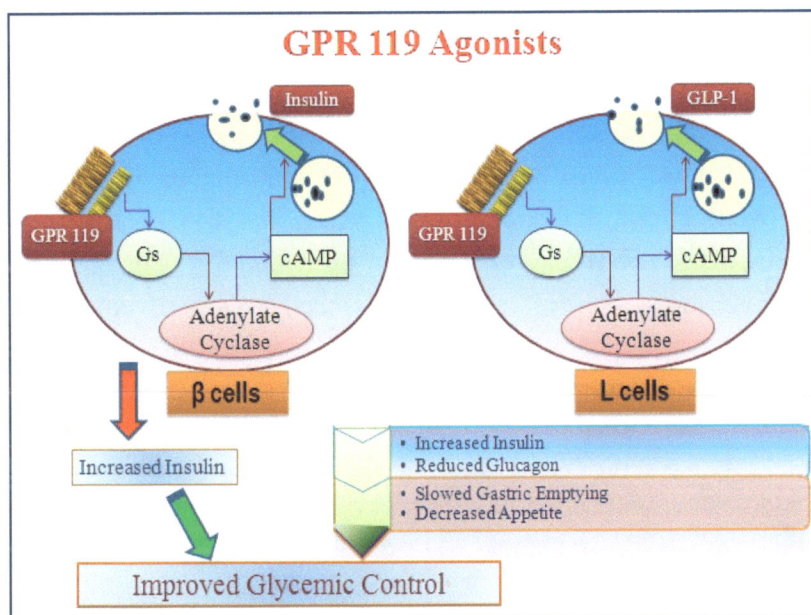

Figure 1: GPR 119 agonists.

The GPR119 agonist AS1907417 is a related modified form of the molecule AS1269574. In animal experiments involving fasted normal mice, a single dose of AS1907417 improved glucose tolerance, without any effect on plasma glucose or insulin levels. Twice-daily doses of AS1907417 for four weeks in diabetic rats could reduce HbA1c levels by 0.8 to 1.5 %. In db/db mice, AS1907417 improved plasma glucose, plasma insulin, lipid profile and increased pancreatic insulin and pancreatic homeobox 1 (PDX-1) mRNA levels. These data demonstrate that it is not only effective in controlling glucose levels, but also in preserving the β-cell function [58]. Another novel GPR119 receptor agonist JNJ-38431055 decreased glucose excursion during an oral glucose tolerance test and increased post-meal total glucagon-like peptide 1 and gastric insulinotropic peptide (GIP) concentrations in humans when compared to placebo. Thus, both the preclinical and clinical data suggest that GPR119 agonists hold a promise as anti-diabetic drugs with glucose lowering as well as β-cell preserving capabilities [59].

G Protein Coupled Receptor-Based Therapies

Glucose metabolism is linked to the sympathetic nervous system (SNS). The α2A-adrenergic receptor (ADRA2A) mediates inhibition of insulin secretion and lipolysis. Alpha (2A) AR mediates adrenergic suppression of insulin secretion and overexpression of alpha2A-adrenergic receptors has been shown to contribute to T2D [60]. Blockade of alpha2A-adrenergic receptors increased insulin secretion in congenic islets in the diabetic Goto-Kakizaki rats [61]. Alpha2A-adrenergic receptor antagonists, yohimbine or efaroxan potentiated glucose-induced insulin release in non-diabetic control rats and produced an improvement in the oral glucose tolerance and potentiated glucose-induced insulin release in type-II diabetic rats [62].

Several experimental studies suggest that the over activity of endocannabinoid system is associated with abdominal obesity and/or diabetes and activation of Cannabinoid type 1 (CB1) receptors has been shown to promote weight gain and associated metabolic changes [63]. Rimonabant, a CB1 receptor blocker reduced body weight, waist circumference, triglycerides, blood pressure, insulin resistance index and C-reactive protein levels, and increased high-density lipoprotein (HDL) cholesterol and adiponectin concentrations in both non-diabetic and diabetic overweight/obese patients. In addition, a 0.5-0.7 % reduction in HbA1c levels was observed in metformin- or sulphonylurea-treated patients with type-2 diabetes and in drug-naïve diabetic patients [64]. Treatment of six-week-old db/db mice with the CB1-specific antagonist SR141716 (10 mg/kg·d) for 3 months significantly improved insulin resistance and lipid abnormalities [65]. Endogenous Cannabinoids have been shown to be generated within β-cells and blockade of CB1R resulted in enhanced insulin receptor signaling and increased β-cell proliferation and mass resulting in reduced blood glucose levels [66].

Newer alpha2A-adrenergic receptor antagonists and CB1 receptor antagonists are being developed and may find a role in the management of type 2 diabetes in the future.

HUMAN GLUCOKINASE (GK) ACTIVATORS

Human glucokinase (GK), also known as hexokinase 4 catalyzes the conversion of glucose to glucose 6-phosphate in both liver and islets. It acts as a glucose-sensing enzyme in the pancreatic β-cells and as a rate-limiting enzyme for hepatic glucose clearance and glycogen synthesis. Activation of this enzyme promotes pancreatic insulin secretion and hepatic glucose uptake and glycogen synthesis, actions which reduce plasma glucose and hence are important in glucose homeostasis. Activating GK mutations cause hyperinsulinemic hypoglycemia and inactivating mutations cause maturity onset diabetes of the young (MODY-2) [67, 68]. Phosphorylation of glucose by glucokinase in β-cell positively affects the rate of ATP production, which leads to closure of potassium channels resulting in insulin secretion. In the liver, at low glucose concentrations, GK is sequestered in the nucleus by binding with its regulatory protein (GKRP) and when glucose concentration increases, GK dissociates from GKRP and is transported into the cytoplasm where it facilitates the uptake of glucose and its conversion into glycogen. Transcription of the GK gene in the liver is induced by insulin and hence hepatic GK activity is insulin-dependent [67, 68]. Half-maximal activity of GK for glucose is approximately 8.0 mM unlike other three hexokinases, which have lower km values (<1.0 mM). Therefore, as the plasma glucose values increase from fasting to postprandial levels, the metabolism of glucose by GK also increases especially so in the liver. High glucose concentration induces GK expression in the β-cell and augments glucose stimulated insulin biosynthesis and release.

About 99 % of the body's GK complement is found within the liver and it plays a critical role during the post prandial state by clearing much of the glucose from the bloodstream along with enhanced glycogen synthesis [67, 69]. In T2D, total β-cell GK activity is thought to be reduced because of loss of β-cell mass. Since hepatic GK is totally insulin dependent its levels are greatly reduced in severe forms of type 2 diabetes. The critical roles of GK in glucose homeostasis coupled with the knowledge that its levels are reduced in T2DM have led researchers to seek actively for the small molecule allosteric activators of GK as a treatment option for T2DM (Fig. **2**). GKAs increase the affinity of GK for glucose by as much as 10-fold and also augment the Vmax by twofold [67- 70].

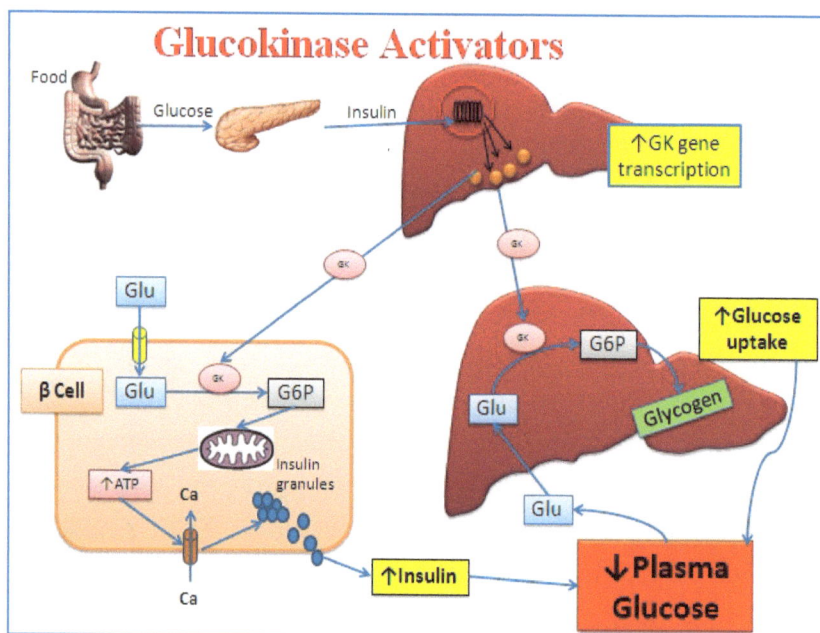

Glu- Glucose; G6P- Glucose 6 phosphate; Ca- Calcium; GK- Glucokinase OR Glucokinase activators

Figure 2: Glucose lowering effects of Glucokinase Activators.

Several glucokinase activators have been developed including piragliatin, compound 14, R1511, AZD1656, AZD6370, compound 6, and ID1101. These molecules demonstrated efficacy in enhancing glucose-stimulated insulin secretion (GSIS) and lowering plasma glucose, but were accompanied by increased risk of hypoglycemia. Continued optimization of structure of glucokinase activators has led to discovery of many new GKAs with enhanced potency and beneficial pharmacokinetic profile. GKA 60 is a potent, soluble glucokinase activator with increased pharmacokinetic half-life and displays an excellent balance of potency, pharmacokinetic and physical properties [71]. It was been postulated that a liver selective activator may offer effective glycemic control with reduced hypoglycemia risk. Carboxylic acid containing series of glucokinase activators with preferential activity in hepatocytes have recently been developed. Several GKAs have entered Phase II clinical trials, and the data on their safety and efficacy profiles are being evaluated.

PROTEIN TYROSINE PHOSPHATASE 1B (PTP1B) INHIBITORS

Protein tyrosine phosphatase 1B (PTP1B) is a member of PTP super family and acts as a negative regulator of insulin and leptin signal transduction. It dephosphorylates membrane-bound and endocytosed insulin receptors in insulin signaling pathway thereby attenuating the tyrosine kinase activity, thus acting as a negative regulator of insulin signaling. This action would be expected to cause insulin resistance. Similarly, it dephosphorylates leptin receptors causing their deactivation and reducing leptin effects [72, 73]. In PTP1B knockout mice, there is increased insulin sensitivity with lower plasma glucose and insulin levels. These mice have lower obesity rates and are protected against weight gain from a high fat diet. In obese diabetic mice, reduction of PTP1B in liver and fat by using antisense oligonucleotides leads to normalization of plasma glucose and insulin. These results indicate that PTP1B inhibition might be effective in reducing both insulin and leptin resistance. Accordingly, inhibition of PTP1B by antisense-based oligonucleotides has been proposed as one of the biological targets in T2D and obesity (Fig. **3**) [74, 75].

PTP1B inhibitors fall into four general classes: difluoromethylene phosphonates, 2-carbomethoxybenzoic acids, 2-oxalylaminobenzoic acids and lipophilic compounds. Difluoromethylene phosphonates possess more potent inhibitory activity against phosphatases in comparison to other inhibitors. PTPs have remained elusive to orally available small-molecule inhibitors that can effectively cross cellular membranes. The low selectivity and poor pharmacokinetic properties of these synthetic inhibitors have led to a continued search for new types of PTP1B inhibitors with improved pharmacological properties [72, 73]. 3-Bromo-4,5-bis-1,2-benzenediol (BDB) derivates with potent PTP1B inhibitor activities have been designed as a series of bromo-retrochalcones based on licochalcone A and E [76]. A series of pyrrolo [2, 3-c] azepine derivatives have also been designed and being evaluated as a new class of inhibitors against protein tyrosine phosphatase 1B (PTP1B) [77]. Orally active inhibitors have also been recently described [78]. But none of them have entered human Phase II trials yet. Although, PTP1B inhibitors are in an early stage of development as anti-diabetic agents, they promise a bright prospect.

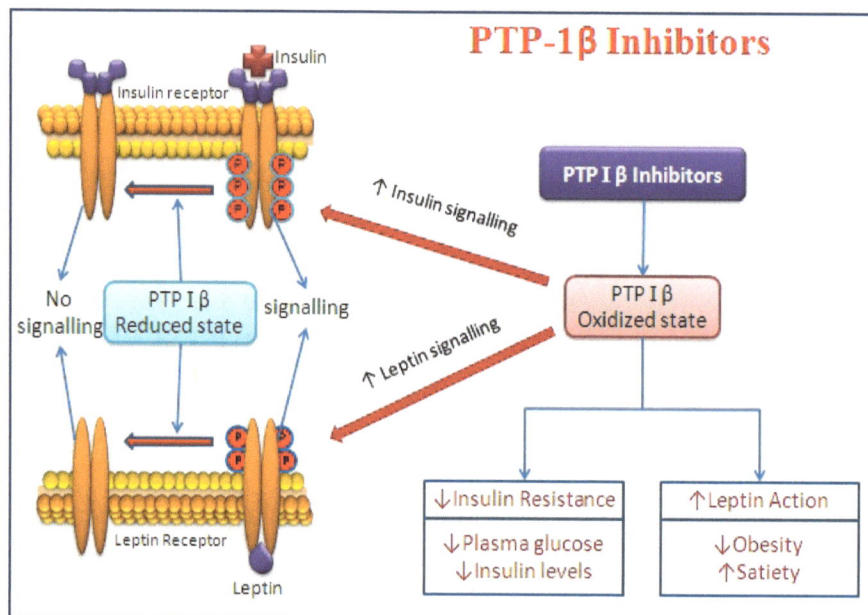

Figure 3: Mechanism of action of Protein Tyrosine Phosphatase 1β Inhibitors.

Low Molecular Weight Protein Tyrosine Phosphatase (LMW-PTP)

Apart from PTP1B, low molecular weight protein tyrosine phosphatase (LMW-PTP) has also been recognized as a negative regulator of insulin signaling. LMW-PTP family also has an oncogenic relevance and acts as a negative regulator of insulin-mediated mitotic signaling. The inhibition of the LMW-PTP is considered to be a good target for the design of new therapeutic drugs in the treatment of T2D and cancer. A new series of chromones has been designed as a new class of highly active LMW-PTP inhibitors. These compounds also inhibit PTP-1B and are active in cellular assays [79]. Further research is required to unearth their actions and properties.

Interleukin-1β Inhibitors

Islet cell inflammation is a shared feature of both types of diabetes mellitus, converging on a common pathway leading to β-cell failure. Interleukin-1 is produced in response to the inflammatory stimuli and mediates various physiologic responses, including inflammatory and immunologic reactions. IL-1β is a member of the interleukin 1 cytokine family and is produced by activated macrophages as a proprotein, which is proteolytically processed to its active form by an enzyme called Caspase. It is a proinflammatory cytokine and being an important mediator of the inflammatory responses it is involved in a variety of cellular activities, including cell proliferation, differentiation, and apoptosis. Interleukin-1β acts as an effector molecule in β-cell inflammation and destruction leading of type 1 diabetes. It also inhibits the function of β-cells and promotes their apoptosis [80]. Intra islet interleukin-1β production has been observed in type 2 diabetes and high glucose levels increase its production and release from β-cells which lead to their reduced function and apoptosis [81]. These findings suggest that it may play a role in glucotoxicity induced β-cell dysfunction seen in T2D. Hence inflammatory process in the islets with IL-1 as a key mediator appears to contribute to the progression of T2D. Further, interleukin-1β receptor antagonist, a naturally occurring competitive inhibitor of interleukin-1 protects human β-cells from glucose-induced functional impairment and apoptosis and the expression of interleukin-1–receptor antagonist is decreased in β-cells obtained from patients with T2D [81 -84]. These findings suggest a role for interleukin-1β in the pathogenesis and progression of T2D and as a potential therapeutic target for preserving β-cell mass and function.

Anakinra

Anakinra is a recombinant human interleukin-1 receptor antagonist. A 13-week intervention study was done to test the efficacy and safety of anakinra, (Kineret) in T2D patients with metabolic control, β-cell function, insulin sensitivity, and inflammatory markers as outcome measures. Anakinra administered subcutaneously once daily for 13 weeks resulted in HbA1c reduction of 0.46 % compared to placebo along with enhanced C-peptide levels, reduced ratio of proinsulin to insulin and reduced levels of interleukin-6 and C-reactive protein [85]. In the follow-up study, patients on anakinra maintained reduced ratio of proinsulin to insulin and the levels of interleukin-6 and C-reactive protein 39-weeks after anakinra was stopped [86]. These results suggest that anakinra has a potential to preserve β-cell function and induce inflammatory remission in patients with T2D. A combined elevation of IL-1β and IL-6 has been shown to predict the risk of developing T2D [87].

Caspase-1/IL-1 β signaling pathway also plays an important role in diabetes-induced retinal pathology. Agents like minocycline and tetracycline which are known to inhibit caspase-1 also inhibit IL-1β production in retinal capillaries. Genetic deletion of the IL-1β receptor in mice protects from diabetes-induced caspase activation and retinal pathology, including capillary loss [88]. IL-1 β receptor antagonism in experimental studies has shown to be cardioprotective during global cardiac ischemia. Administration of anakinra in experimental models of acute myocardial infarction in rats significantly reduced the remodeling process by inhibiting cardiomyocyte apoptosis. This benefit was observed in two different experimental animal models of acute myocardial infarction (MI) when it was administered within 24 hours of MI [89].

Gluconeogenesis Inhibitors

Gluconeogenesis results in the generation of glucose from non-carbohydrate carbon substrates such as lactate, glycerol, and gluconeogenic amino acids. T2D is characterized by excessive endogenous glucose production leading to hyperglycemia. Fasting hyperglycemia in T2D is determined mainly by hepatic glucose production resulting from gluconeogenesis. Gluconeogenesis occurs mainly in the liver (90 %) and kidneys (10 %). Pyruvate, lactate, glycerol, amino acids and citric acid cycle intermediates are substrates for gluconeogenesis. Free fatty acids also induce gluconeogenesis, although they are not substrates for it. The rate of gluconeogenesis is determined by the action of a rate controlling enzyme, fructose-1, 6-bisphosphatase, which is regulated by cAMP and its phosphorylation. Factors regulating the activity of the gluconeogenesis pathway do so mainly by altering the activity or expression of key enzymes. Fructose 1, 6 bisphosphatase, the key hepatic enzyme in gluconeogenesis is an important target for gluconeogenesis inhibitors [90]. FBPase is also thought to play an important role in regulating glucose sensing and insulin secretion of β-cells [91]. FBPase inhibitors are hence a promising target for diabetes treatment. Inhibitors of FBPase are different from current OADs which do not reduce gluconeogenesis effectively and hence can fulfill one of the unmet needs in T2DM pharmacotherapy.

AMP binding site of fructose 1, 6-bisphosphatase is the main target for potent gluconeogenesis inhibitors. By using a structure-based drug design strategy, a series of compounds that mimic AMP but bear little structural resemblance have been discovered. MB06322 is the best characterized of these compounds [92, 93]. Oral administration of MB0632 (CS-917) to zucker diabetic fatty rats resulted in a dose-dependent decrease in FPG levels [94]. CS-917 reduced the rate of gluconeogenesis and hepatic glucose production in diabetic rodents and fasting plasma glucose decreased mainly because of reduced hepatic glucose release and lactate uptake from liver. It reduced both fasting and postprandial hyperglycemia after meal loading in non-obese type 2 diabetic Goto-Kakizaki rats [95, 96]. Furthermore, it was not associated with any weight gain or hypoglycemia in animal studies.

MB07803

MB07803 is a second generation FBPase inhibitor with improved pharmacokinetics as compared to MB06322. MB07803 is an oral product like CS-917 and was discovered using our NuMimetic technology. While MB07803 is structurally different from CS-917 and may offer certain pharmacological advantages it targets the same binding site on the FBPase enzyme [97]. In a Phase II clinical trial of MB07803 involving 105 patients with type 2 diabetes, MB07803 at 200 mg resulted in a statistically significant and clinically meaningful placebo-adjusted reduction in FPG from baseline of -28.9 mg/dL at day 28. It was safe and well tolerated with 94 % of the patients completing the study and there were no patient withdrawals due to drug-related adverse events (AEs). Future trials of MB07803, will address whether this novel class of anti-diabetic agents can provide safe and long-term glycemic control. If approved, FBPase inhibitors could provide an additional therapeutic option for the large number of patients with diabetes including those intolerant to metformin.

IMEGLIMIN

Imeglimin is a tetrahydrotriazine compound which belongs to a new class of drugs – the glimins. It has unique pharmacological properties that are different from other OADs and has been designed to effectively treat the underlying metabolic defects in patients with T2D. It acts on the liver, muscle and pancreatic β-cells and targets the key defects of T2D. It is an oxidative phosphorylation blocker in the mitochondria that has three main actions –inhibition of hepatic glucose production, increase the muscle glucose uptake and increase pancreatic glucose-dependent insulin secretion. Imeglimin also improves both β-cell function and survival. Insulin secretion observed with imeglimin is glucose dependent like that seen with GLP-1 analogues, safeguarding against the risk of hypoglycemia. It is also thought to act as an indirect AMPK activator. Imeglimin is under investigation for the treatment of type 2 diabetes [98, 99].

Imeglimin is being developed by Poxel and is in Phase II clinical development for T2D. It has achieved its clinical proof of concept in Phase IIa trials. Imeglimin has been tested in two animal models of T2D, the GK and the STZ rat models. In these models, after 5 and 8 weeks of treatment, Imeglimin improved FPG, PPG, HbA1c and glucose tolerance. In animal models it also increases glucose uptake [98]. In a phase II study, Imeglimin (2000 mg qd and 1000 mg bid) was compared with metformin over 4 weeks on glucose area under curves (AUC) during an OGTT. Baseline-adjusted changes in the OGTT area under curves (AUC) were -33 % for imeglimin bid, -30 % for metformin and -10 % for imeglimin qd. In a related phase II study, two daily doses of imeglimin (500 mg bid, 1500 mg bid) were compared with placebo and metformin (850 mg bid) over 8 weeks and AUC glucose during a prolonged meal test, fasting plasma glucose (FPG) and HbA1c were recorded. The changes in AUC 0-6h were significantly different from placebo for both imeglimin 1500 mg and metformin. Imeglimin 1500 mg reduced FPG and HbA1c. 500 mg dose of imeglimin had a very limited efficacy. Tolerability profile of imeglimin was encouraging in phase 2 trials [99, 100].

In summary, imeglimin displays a superior benefit: risk profile compared to metformin and an encouraging tolerability profile in T2D patients. Future studies in a wider patient population will determine its utility in diabetes and whether it is suitable for combination with other classes of anti-diabetic agents.

5' Adenosine Monophosphate-Activated Protein Kinase Activators

AMPK (5' adenosine monophosphate-activated protein kinase) is a protein kinase which plays a key role in regulating energy metabolism in organs like liver, adipose tissue, skeletal muscle and hypothalamus. Liver AMPK activation leads to inhibition of gluconeogenesis and hepatic glucose production. In the muscle tissue, AMPK activation increases fatty acid oxidation resulting in decreased intramyocyte lipid accumulation and improved insulin sensitivity. In the hypothalamus, AMPK activation stimulates food intake. AMPK is thought to act as an integrator of regulatory signals reflecting systemic and cellular energy status. AMPK-mediated glucose uptake in muscle cells from patients with T2D has been shown to be intact. AMPK is a key regulator of glucose, lipid and energy metabolism and activation of AMPK in the liver and muscle in diabetic patients is expected to result in a spectrum of beneficial metabolic effects [101, 102].

A number of hormones and pharmacological agents have been reported to activate AMPK indirectly. Exercise, adiponectin, IL-6, leptin, resveratrol, natural alkaloids like bitter melon extracts and berberine modulate its activity [103]. Herbal plant, Gynostemma pentaphyllum contains Damulins A and B which activate AMPK [104]. Metformin and pioglitazone act indirectly to activate AMPK. PPARγ activators, rosiglitazone and pioglitazone, activate AMPK by increasing cellular AMP/ATP ratio. AICAR (5-Aminoimidazole-4-carboxamide-1-β-Dribonucleoside), an analogue of natural activator AMP, activates AMPK through direct binding followed by allosteric modification. Metformin activates AMPK by an unknown mechanism. Adiponectin, an insulin sensitizing adipokine also acts to increase AMPK activity. Physical activity leads to AMPK activation in skeletal muscle, liver and adipose tissue. Improvement in glycemic control and insulin sensitivity due to physical activity is at least partly mediated through AMPK activation [103, 105].

AMPK is expressed in specific nuclei of the hypothalamus and its activation increases food intake [106]. The ideal AMPK activators should activate AMPK at low concentration and be effective in target organs such as the liver and skeletal muscle but not the hypothalamus. Tissue-specific pharmacological activation of AMPK could be achieved through isoform-specific activation of AMPK. Small molecule AMPK activator, A-769662 is mainly targeted to the liver [106]. Effects of AICAR infusion in T2D patients are also largely restricted to the liver. AICAR showed promise in preclinical models as an anti-diabetic drug but failed in phase I clinical trials. Development of potent and selective AMPK activators is underway and preliminary proof of concept in preclinical models suggests a great hope for therapeutic use in diabetes (Fig. **4**).

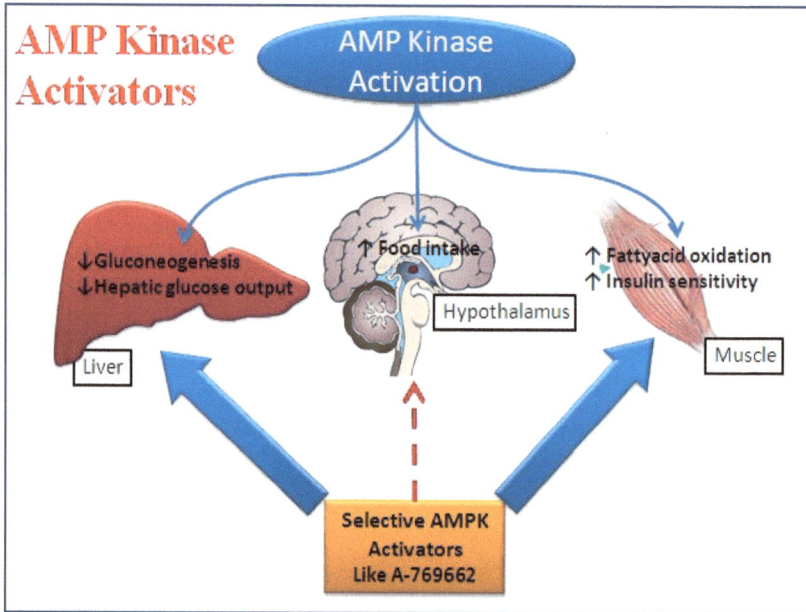

Figure 4: Beneficial metabolic effects of AMP Kinase Activators in diabetes.

11β-Hydroxysteroid Dehydrogenase Type 1 Inhibitors

Obesity, T2D and Cushing's syndrome share many phenotypic features and there is enough evidence for a plausible role of endogenous glucocorticoids in the pathogenesis of T2D. 11β-hydroxysteroid dehydrogenase type 1 (11β-HSD1) is a NADPH-dependent enzyme which catalyzes the conversion of inactive cortisone to active cortisol. 11β-HSD1 is highly expressed in liver, adipose tissue and brain where its effects result in amplification of local glucocorticoid action. In transgenic mice, over expression of 11β-HSD1-selectively in adipose tissue leads to development of T2D and visceral obesity, whereas 11β-HSD1 knockout mice display normoglycemia and lean body phenotype with resistance to weight gain on a high fat diet. 11β-HSD1 activity has been shown to be elevated in adipose tissue of obese rodents and humans. 11β-HSD1 mediated intracellular cortisol production may have a pathogenic role in T2D and 11β-HSD1 inhibition has been explored as a potential therapy for T2D and metabolic syndrome [100, 101]. Intracellular cortisol production by11β-HSD1 can have detrimental effects on glycemic levels, blood pressure and lipid levels. Its activity also leads to increased glucose production in liver and insulin resistance by inhibiting glucose uptake and disposal of glucose in muscle and adipose tissues [107-110]. Given these effects, 11β-HSD1 inhibition can be expected to be beneficial for T2D patients (Fig. **5**).

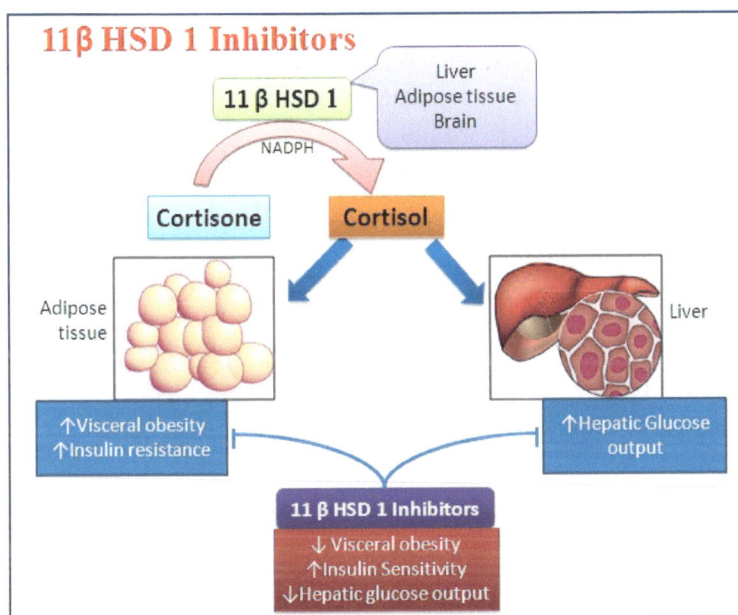

Figure 5: 11β-Hydroxysteroid Dehydrogenase Type 1 Inhibitors on diabetes and obesity.

INCB13739 & MK-0916

A number of 11β -HSD1 inhibitors have been designed and discovered but most of them failed to reach Phase III clinical trials. INCB13739 from Incyte Inc. is a novel and selective 11β -HSD1 inhibitor which showed promise in animal and preclinical studies is in Phase IIB clinical trials [111, 112]. In a Phase II dose-ranging study, 302 patients with T2D on metformin monotherapy were randomized to receive INCB13739 or placebo once daily for 12 weeks. After 12 weeks, 200 mg of INCB13739 resulted in significant reductions in A1C (-0.6 %), fasting plasma glucose (-24 mg/dl), and homeostasis model assessment-insulin resistance (HOMA-IR) (-24 %) compared with placebo. It also reduced body weight and improved lipid parameters [113]. Another selective 11β -HSD1 inhibitor MK-0916 resulted in modest improvements in A1C, body weight and blood pressure when compared to placebo in patients with T2D and metabolic syndrome. No significant improvement in FPG was observed [114]. In another phase II study in MK-0916 did not produce clinically significant improvements in BP endpoints, LDL-C, and body weight [115]. AMG 221, a selective 11β-HSD1 inhibitor is in early stage of clinical development [116].

Selective inhibition of 11β-HSD1 and the reduction in cortisol levels in key metabolic tissues have the potential to simultaneously target multiple cardiovascular risk factors in patients with T2D and obesity.

Peroxisome Proliferator-Activated Receptor (PPAR) Dual Agonists

Peroxisome proliferator-activated receptors (PPARs) are a group of nuclear receptor proteins that function as ligand-activated transcription factors modulating the expression of various target genes. They are members of the steroid hormone nuclear receptor family. PPARs like other members of nuclear receptor family form heterodimer with the 9-cis-retinoic acid receptor (RXR). There are 3 PPAR subtypes – PPAR α, γ, and δ. PPAR-α is expressed mainly in skeletal muscle and liver, where it is involved in lipoprotein metabolism and fatty acid oxidation. PPAR-α agonists increase fatty acid oxidation, increase HDL, decrease circulating triglycerides and have anti atherosclerotic activity. They also reduce inflammation and stabilize atheromatous plaques. PPAR-α is the molecular target of fibrates. PPAR-γ is expressed mainly in fat and muscle. PPAR-γ plays an important role in adipogenesis, lipid metabolism, glucose control and regulation of adipocyte differentiation and function. PPAR-γ agonists improve insulin sensitivity by increasing peripheral adipose tissue lipogenesis while reducing visceral and hepatic fat content. They cause redistribution of fat from visceral to peripheral tissues and this results in improved insulin sensitivity. Reduced hepatic fat content in turn leads to reduced hepatic glucose production. Table **3** summarizes all the three classes of PPAR agonists. Thiazolidinediones were serendipitous identified drug screening as a class of drugs that improve insulin sensitivity by binding to and activating PPAR-γ. The insulin-sensitizing drugs rosiglitazone and pioglitazone in current clinical use are PPAR-γ agonists.

Table 3: Summary of PPAR agonists.

	Major Effects	**Tissue Expression**	**Agonists**
PPAR-α	↑ fatty acid oxidation ↑ HDL ↓triglycerides ↓ inflammation Anti atherosclerotic properties	Skeletal muscle and liver	Fibrates- to manage dyslipidemia
PPAR-γ	Fat redistribution from visceral sites to periphery ↓ visceral fat ↑ insulin sensitivity ↓ hepatic glucose production ↓ plasma glucose	Fat and muscle	Glitazones- to reduce insulin sensitivity and reduce blood glucose in diabetes
PPAR-δ	Regulate body's energy fuel preference from glucose to fat Regulate energy metabolism ↑ HDL ↓triglycerides ↑oxidative slow-twitch fibers Probable role in cancer	Widespread	Fatty acids or eicosanoids

The third isotype is PPAR-δ, which is expressed in essentially all cells throughout the body. PPAR-δ is thought to play an important role in cell biology and energy metabolism, and has also been implicated in inflammation control. PPAR-δ agonists regulate body's energy fuel preference from glucose to fat and reduce blood glucose levels and are being investigated for obesity and diabetes. PPAR-γ (PPAR-γ) agonists TZDs (pioglitazone and rosiglitazone) improve insulin sensitivity and reduce blood glucose values. PPAR-α agonists like fibrates promote fatty acid oxidation and reduce hyperlipidemia. To achieve combined benefits of PPAR alpha on lipid metabolism and PPAR-γ on insulin sensitivity, a number of PPAR α/γ dual agonists have been developed. Dual PPAR agonists combine the lipid benefit of PPAR-α agonists, such as fibrates, with the glycemic advantages of the PPAR-γ agonists, such as the thiazolidinediones (Fig. **6**). These therapeutic benefits can be achieved without excessive weight gain [117, 118]. Dual PPAR-α/γ agonists muraglitazar, ragaglitazaar and tesaglitazar reached phase III clinical stage and later were discontinued due to serious side effects.

PPAR Dual Agonists

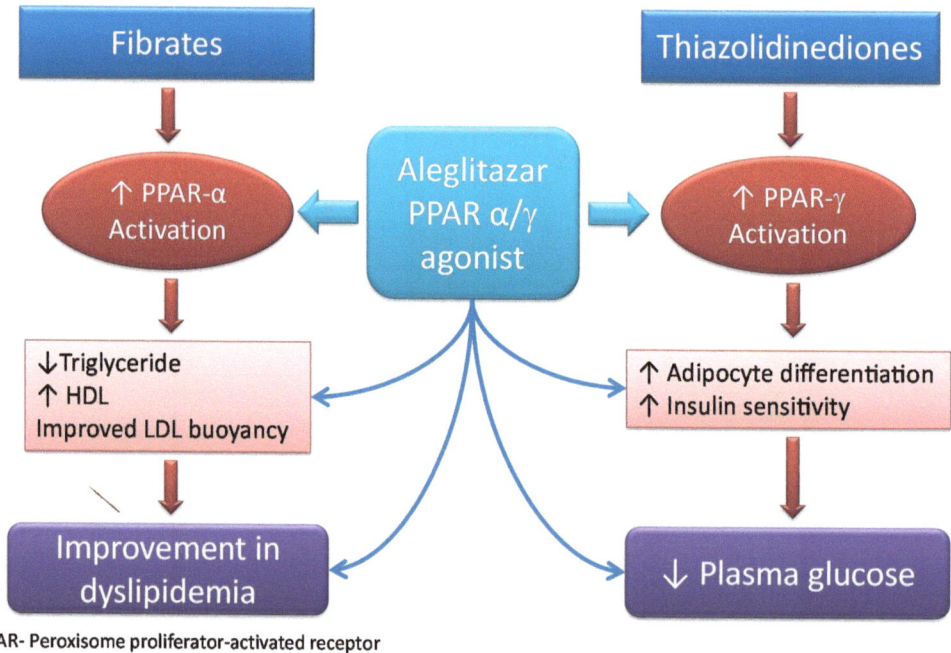

PPAR- Peroxisome proliferator-activated receptor

Figure 6: Peroxisome Proliferator Activated Receptor dual agonists.

Aleglitazar

Aleglitazar is a balanced dual PPAR-α/γ agonist molecularly designed to optimize glycemic and lipid benefits, and minimize weight gain and edema [119]. In a phase II, dose ranging study (SYNCHRONY) Aleglitazar significantly reduced baseline HbA1c in comparison to placebo in a dose-dependent manner ranging from –0·36 % to –1·35 %. Edema, hemodilution, and weight gain also occurred in a dose-dependent manner [120]. Aleglitazar 0.03 mg/kg per day reduced triglyceride levels increased HDL, apolipoprotein A-I in a primate model of the metabolic syndrome. It also improved insulin sensitivity and reduced body weight in this study [121]. Aleglitazar has been shown to cause a reversible dose-dependent decrease in GFR attributable to intrarenal hemodynamic effects. In a randomized study on healthy volunteers (n=44) aged 40-65 years with normal renal function, aleglitazar (150 mcg qd) for 4 weeks did not change the effect size of alterations in renal function induced by low-dose aspirin alone [122]. Phase III studies are underway to assess its efficacy and side-effect profile.

Chiglitazar

Chiglitazar is another novel PPARalpha/gamma dual agonist which improved insulin resistance and dyslipidemia in obese rats. It is in phase II clinical trials [123]. A novel non thiazolidinedione PPAR/dual agonist, CG301360 improved insulin sensitivity, enhanced fatty acid oxidation and glucose uptake and also reduced pro-inflammatory gene expression in db/db Mice. In experimental studies, CG301269 selectively stimulated the transcriptional activities of PPARα and PPARγ, enhanced fatty acid oxidation and ameliorated insulin resistance and hyperlipidemia. In db/db mice, CG301269 reduced inflammatory responses and fatty liver, without body weight gain [124].

ANTI OBESITY DRUGS WITH GLUCOSE LOWERING POTENTIAL

QNEXA

QNEXA is a novel weight loss therapy that combines low doses of two agents, phentermine and topiramate. It is also being studied for its positive effect on patients with OSA or the obstructive sleep apnea. Phentermine and Topiramate are the two major components of QNEXA. Phentermine, a central norepinephrine-releasing drug, suppresses hunger. Topiramate is an antiepileptic drug which has weight-loss properties. Topiramate-induced weight loss mainly results from increased satiety, but other contributing factors include increased taste aversion, increased energy expenditure, and decreased caloric intake. Phentermine and Topiramate suppress appetite through distinct and complementary mechanisms (decreased hunger and increased satiety) and hence the combination is being recommended as a suitable agent for weight reduction and obesity treatment [125, 126].

Apart from weight loss, QNEXA improved the glycemic profiles of patients in phase II and III clinical trials. In OB 202, a 28 wks Phase II clinical trial, QNEXA reduced HbA1c by 1.2 %, from 8.7 % to 7.5 % compared to 0.6 % in the placebo group (p<0.001). FPG levels were reduced in the QNEXA arm from 174.7 mg/dL to 141.9 mg/dL. In OB-303, a phase III clinical trial QNEXA resulted in improvements in fasting insulin levels, insulin resistance as measured by HOMA-IR and insulin sensitivity during oral glucose tolerance testing. QNEXA demonstrated a benefit in retardation of progression from IFG/IGT to T2D in Phase III studies. It also improved other obesity-related co morbidities, such as BP, lipid co morbidities and cardiovascular inflammatory markers [126 -129]. QNEXA represents a significant advance in medical therapy of obesity and management of weight-related co morbidities and is pending FDA approval.

Lorcaserin

Lorcaserin is a selective serotonin (5-HT) 2C receptor agonist. Activation of 5-HT 2C receptors within the hypothalamus increases proopiomelanocortin (POMC) production leading to enhanced satiety. Lorcaserin has 100:1 affinity for 5-HT 2C *versus* other receptors. It has been shown that weight loss by lorcaserin is due to reduced energy intake and not due to enhanced energy expenditure [130]. Lorcaserin also has beneficial effects on risk factors for T2D and cardiovascular disease, including BP, heart rate, and levels of LDL cholesterol [131]. In addition to weight loss, Lorcaserin leads to improvement in glycemic control in patients with T2D. In a 52-week randomized control trial of 604 patients with T2D, HbA1c decreased 0.9 % with lorcaserin 10mg bid, 1.0 % with lorcaserin qd, and 0.4 % with placebo and FPG decreased by 27.4 mg/dl, 28.4 mg/dl and 11 mg/dl respectively [132]. Recently, lorcaserin secured FDA approval for the treatment of obesity in adults with a BMI ≥ 30 kg/m2 or adults with a BMI of 27 kg/m2 or greater with weight related co morbidities.

STEM CELL BASED THERAPIES

Although reduced β-cell mass is a hallmark of T1D, T2D is also associated with β-cell dysfunction and β-cell loss. At the diagnosis of T2D, β-cell numbers are reduced by 40 % to 60 %. Most people with insulin resistance who progress to the diabetic state have reduced β-cell mass. β-cell mass is substantially reduced in long standing T2D and pancreas transplantation done then often restores normal glucose levels. Advances in stem cell biology have the potential of β-cell restoration in both forms of diabetes. There is also a great interest in mesenchymal stromal cells to promote islet cell regeneration, which may locally also increase growth factors and transcription factors that are helpful in β-cell regeneration. Stem cells might also lead to innovative solutions to the problems of obesity and insulin resistance. β-cell regeneration using stem cell therapy may prove to be an effective modality for T2D [133].

The metabolic effects of combination of autologous bone marrow stem cell transplantation (BMT) and hyperbaric oxygen treatments on T2D were studied in 31 adult patients. HbA1c reduced from 8.7 % to 7.1 % by 30 days after BMT with no further improvement. At 90 days after the combined therapy, C-peptide increased significantly compared with baseline but reduced thereafter. All patients had insulin and/or oral hypoglycemic drugs reduced to different levels at least for some period of time. Overall, the improvement in pancreatic β-cell function was transient [134]. Another study evaluated autologous bone marrow-derived stem cell transplantation (SCT) in 10 patients with T2D for more than five years, who were on insulin due to failure of triple OAD. Seven of the ten patients were responders and in them, there was a reduction in the insulin requirements by 75 % from the baseline. Three patients could discontinue insulin completely, although it was short-lived in one of them. Mean HbA1c reduction was 1 % at six months. Both the fasting and glucagon-stimulated C-peptide level showed significant improvements in the entire group after six months [135].

Therapeutic benefit of human placenta-derived mesenchymal stem cells (PD-MSC) was studied in 10 long standing T2D patients with beta cell dysfunction, high insulin doses and poor glycemic control. Patients were given three intravenous infusions of PDSC at one-month intervals. The daily mean dose of insulin decreased from 63.7 to 34.7 IU ($P<0.01$), and the C-peptide level increased from 4.1 ng/mL to 5.6 ng/mL ($P<0.05$) at three months. In four of the patients, insulin doses reduced by more than 50 % after three months of infusion [136]. In summary, stem cell approaches appear to be a safe and effective modality for improvement of beta-cell function in patients with T2D. Prospective large-scale studies are warranted to verify these observations.

BARIATRIC AND METABOLIC SURGERIES

Prevalence of diabetes increases with increasing body mass index (BMI) and nearly 50 % of those diagnosed with T2D are obese. Weight control is the key to successful T2D management. Weight loss has been shown to be effective in preventing and treating T2D. Conventional treatment, including lifestyle modification and pharmacotherapy produce small improvements in weight whereas bariatric surgery has been shown to provide durable weight loss.

Bariatric procedures are classified as restrictive, malabsorptive, or combined procedures, explaining the mechanism of weight loss. Restrictive procedures include laparoscopic adjustable gastric banding (LAGB) and vertical banded gastroplasty (VBG). These procedures reduce the volume of the stomach to decrease food intake and induce early satiety. Malabsorptive procedures, such as the biliopancreatic diversion (BPD), shorten the small intestine to decrease nutrient absorption. Combined procedures, such as the Roux-en-Y gastric bypass (RYGB), incorporate both restrictive and malabsorptive elements. Roux-en-Y gastric bypass surgery is the current gold standard treatment for severe obesity [137].

The secretion of anorexigenic and orexigenic gut peptides is altered by BPD and RYGB procedures favoring weight loss. Levels of GLP-1, Peptide YY (PYY), and ghrelin are reduced in obese patients and even further in diabetic patients. Bariatric procedures such as BPD and RYGB that expedite nutrient delivery to the distal ileum, increase GLP-1 and PYY levels. Ghrelin levels are also reduced by some procedures. These alterations contribute to weight loss by interaction with appetite centers within the hypothalamus to decrease appetite [138, 139]. Bariatric surgeries reduce mean excess body weight by 50-70 % depending on the procedure and reduce the body mass index by 10 to 15 kg/m2 and weight by 20 to 50 kgs. Available evidence from randomized trials and observational studies suggests that bariatric surgery may be associated with a 50 % to 80 % rate of diabetes resolution. A recent systematic review and meta-analysis of 621 bariatric surgery studies, including >135,000 patients revealed that diabetes "improved or resolved" in over 80 % of patients [139]. Longer duration of diabetes (>10 years), poor preoperative glycemic control, and preoperative insulin use are poor predictors of diabetes resolution after surgery, whereas preoperative treatment only with oral anti diabetic agents, smaller preoperative waist circumference and shorter duration of diabetes are associated with increased probability of diabetes resolution. Bariatric surgeries are currently indicated for obese patients with BMI >40 kg/m2 or for patients with a serious co morbidity such as type 2 diabetes and BMI >35 kg/m2 [139 -141]. Surgery is also appropriate in the treatment of patients with T2D and obesity not achieving the recommended therapeutic targets with medical therapies, particularly in the presence of another major obesity associated co-morbidity [142].

Metabolic surgery is now defined as any modification of the gastrointestinal (GI) tract, where rerouting the food passage seems to improve T2DM, based on mechanisms that are weight loss independent. This new frontier of bariatric/metabolic surgery includes the application of conventional bariatric procedures such as RYGB, BPD and sleeve gastrectomy. It also includes the introduction of new procedures such as mini-gastric bypass, ileal interposition with sleeve gastrectomy, ileal interposition with diverted sleeve gastrectomy, duodenal jejunal bypass and stomach- and pylorus-preserving BPD, which are designed with the aim of achieving specific metabolic effects regardless of the effect on weight loss [143].

In ileal interposition, a 10–20 cm portion of an intact ileum is transposed into the proximal region of the small intestines. Early delivery of nutrient-rich chyme to the ileum occurs following ileal interposition. This has been shown to reliably reduce the food intake and body weight in animal studies [144]. Animal studies of ileal transposition show exaggerated release of the GLP-1 [145]. In a recent study on 202 patients with T2D and BMI below 35, laparoscopic ileal interposition into the jejunum (JII-SG) or into the duodenum (DII-SG) associated with sleeve gastrectomy were associated with a mean BMI reduction from 29.7 to 23.5 kg/m^2. HbA1c was decreased from 8.7 to 6.2 % after the JII-SG and to 5.9 % following the DII-SG. HbA1c below 7 % was seen in 89.9 % of the patients and below 6.5 % in 78.3 %. Overall, 86.4 % of patients were off anti-diabetic medications [146].

Duodenal-jejunal bypass (DJB) involves a roux-en-y bypass of the duodenum and a segment (30-50cm) of proximal jejunum bypassing the stomach which is spared.

It is currently being investigated in human trials. In one study DJB had a moderate effect with HbA1c reduction from 9.3 % to 7.7 % at 12 months after surgery, a 20 % decrease in the use of diabetes medications and remission of diabetes in 20 % of subjects. DJB has the potential to replicate the safety profile of the Roux-en-Y gastric bypass while maintaining at least partial efficacy [147].

SUMMARY

The ideal antihyperglycaemic drug for diabetes is still an elusive goal. All possible avenues to develop drugs to address different pathophysiological aspects of Type 2 diabetes are warranted. It is not only pertinent as to how many of these novel therapies are effective, but also their safety and cardiovascular effects. The limitations, side effects and the difficulties in achieving sustained glycemic control with established anti-diabetic therapies have intensified the quest for new medications for diabetes. Long acting GLP-1 analogues appear to be the future of incretin based parenteral therapies. DPP-4 inhibitors have so far been very successful and the future appears bright for them. Newer gliptins are lining up to be approved by regulatory authorities. SGLT2 inhibitors are on the threshold of being approved for clinical use. Mega clinical trials and longer studies are necessary for bariatric surgery/metabolic surgery before they are accepted as one of the therapies for diabetes.

ACKNOWLEDGEMENTS

Declared none.

CONFLICT OF INTEREST

The authors confirm that this chapter content has no conflicts of interest.

REFERENCES

[1]   DeYoung MB, MacConell L, Sarin V, Trautmann M, Herbert P. Encapsulation of exenatide in poly-(D,L-lactide-co-glycolide) microspheres produced an investigational long-acting once-weekly formulation for type 2 diabetes. Diabetes Technol Ther 2011 Nov; 13(11): 1145-54

[2]   Tomkin GH. Albiglutide, an albumin-based fusion of glucagon-like peptide 1 for the potential treatment of type 2 diabetes. Curr Opin Mol Ther 2009 Oct;11(5): 579-88.

[3]   Bush MA, Matthews JE, De Boever EH, *et al.,* (2009) Safety, tolerability, pharmacodynamics and pharmacokinetics of albiglutide, a long-acting glucagon-like peptide-1 mimetic, in healthy subjects. Diabetes Obes Metab 11: 498–505.

[4]   Dong JZ, Shen Y, Zhang J, Tsomaia N, Mierke DF, Taylor JE. Discovery and characterization of taspoglutide, a novel analogue of human glucagon-like peptide-1, engineered for sustained therapeutic activity in type 2 diabetes. Diabetes Obes Metab

[5]   Jimenez-Solem E, Rasmussen MH, Christensen M, Knop FK. Dulaglutide, a long-acting GLP-1 analog fused with an Fc antibody fragment for the potential treatment of type 2 diabetes. Curr Opin Mol Ther 2010 Dec; 12(6): 790-7.

[6]   Umpierrez G, Blevins T, Rosenstock J, Cheng C, Bastyr E, Anderson J. The effect of LY2189265 (GLP-1 analogue) once weekly on HbA(1c) and beta cell function in uncontrolled type 2 diabetes mellitus: the EGO study analysis. Diabetologia 2009; 52: S592011 Jan;13(1): 19-25.

[7]   Drucker DJ, Buse JB, Taylor K, *et al.,* DURATION-1 Study Group. Exenatide once weekly *versus* twice daily for the treatment of type 2 diabetes: a randomized, open-label, non-inferiority study. Lancet 2008; 372: 1240-1250

[8]   Aroda VR, DeYoung MB. Clinical implications of exenatide as a twice-daily or once-weekly therapy for type 2 diabetes. Postgrad Med 2011 Sep;123(5): 228-38.

[9]   St Onge EL, Miller SA Albiglutide: a new GLP-1 analog for the treatment of type 2 diabetes. Expert Opin Biol Ther 2010;10: 801–806

[10]   Madsbad S, Kielgast U, Asmar M, Deacon CF, Torekov SS, Holst JJ. An overview of once-weekly glucagon-like peptide-1 receptor agonists--available efficacy and safety data and perspectives for the future. Diabetes Obes Metab 2011 May; 13(5): 394-407.

[11]   Tzefos M, Harris K, Brackett A. Clinical efficacy and safety of once-weekly glucagon-like peptide-1 agonists in development for treatment of type 2 diabetes mellitus in adults. Ann Pharmacother 2012 Jan; 46(1): 68-78.

[12]   Raz I, Fonseca V, Kipnes M, *et al.,* Efficacy and safety of taspoglutide monotherapy in drug-naive type 2 diabetic patients after 24 weeks of treatment: results of a randomized, double-blind, placebo-controlled phase 3 study (T-emerge 1). Diabetes Care 2012 Mar;35(3): 485-7.

[13]   Thibaudeau K, Robitaille M, Wen S *et al.* CJC-1134-PC: an exendin-4 conjugate with extended pharmacodynamic profiles in rodents. Diabetes 2006; 55: A103.

[14]   Wang M, Kipnes M, Matheson S *et al.* Safety and pharmacodynamics of CJC-1134-PC, a novel GLP-1 receptor agonist, in patients with type 2 diabetes mellitus: a randomized, placebo-controlled, double-blind, dose escalation study. Diabetes 2007; 56: A133.

[15]   Baggio LL, Huang Q, Cao X, Drucker DJ. An albumin-exendin-4 conjugate engages central and peripheral circuits regulating murine energy and glucose homeostasis. Gastroenterology 2008; 134: 1137–1147.

[16]   Wang M, Matheson S, Picard J, Pezzullo J, Ulich T. PC-DAC (TM): exendin-4 (CJC-1134-PC) significantly reduces HbA1c and body weight as an adjunct therapy to metformin: two randomized, double-blind, placebo-controlled, 12 week, phase II studies in patients with type 2 diabetes mellitus. Diabetes 2009; 58: A148.

[17]   Garber AJ. Long-acting glucagon-like peptide 1 receptor agonists: a review of their efficacy and tolerability. Diabetes Care 2011 May; 34 Suppl 2:S279-84.

[18]   Christensen M, Knop FK. Once-weekly GLP-1 agonists: How do they differ from exenatide and liraglutide? Curr Diab Rep 2010 Apr; 10(2): 124-32.

[19] Pinelli NR, Hurren KM. Efficacy and safety of long-acting glucagon-like peptide-1 receptor agonists compared with exenatide twice daily and sitagliptin in type 2 diabetes mellitus: a systematic review and meta-analysis. Ann Pharmacother 2011 Jul; 45(7-8): 850-60.

[20] Christensen M, Knop FK, Holst JJ, Vilsboll T. Lixisenatide, a novel GLP-1 receptor agonist for the treatment of type 2 diabetes mellitus. IDrugs 2009 Aug;12(8): 503-13. Review.

[21] Hunter K, Holscher C. Drugs developed to treat diabetes, liraglutide and lixisenatide, cross the blood brain barrier and enhance neurogenesis. BMC Neurosci 2012 Mar 23; 13(1): 33.

[22] Ratner RE, Rosenstock J, Boka G; DRI6012 Study Investigators. Dose-dependent effects of the once-daily GLP-1 receptor agonist lixisenatide in patients with Type 2 diabetes inadequately controlled with metformin: a randomized, double-blind, placebo-controlled trial. Diabet Med 2010 Sep; 27(9): 1024-32.

[23] Fonseca VA, Alvarado-Ruiz R, Raccah D, Boka G, Miossec P, Gerich JE; on behalf of the EFC6018 GetGoal-Mono Study Investigators. Efficacy and Safety of the Once-Daily GLP-1 Receptor Agonist Lixisenatide in Monotherapy: A randomized, double-blind, placebo-controlled trial in patients with type 2 diabetes (GetGoal-Mono). Diabetes Care 2012 Jun; 35(6): 1225-1231

[24] Barnett AH. Lixisenatide: evidence for its potential use in the treatment of type 2 diabetes. Core Evid 2011; 6: 67-79

[25] Pan CQ, Buxton JM, Yung SL, *et al.* Design of a long-acting peptide functioning as both a glucagon-like peptide-1 receptor agonist and a glucagon receptor antagonist. J Biol Chem. 2006;281: 12506-12515

[26] Claus TH, Pan CQ, Buxton JM *et al.* Dual-acting peptide with prolonged glucagon-like peptide-1 receptor agonist and glucagon receptor antagonist activity for the treatment of type-2 diabetes. Journal of Endocrinology 2007; 192: 371–380.

[27] Fosgerau, K, Skovgaard, M., Larsen, S.A., *et al.,* Combination of long-acting insulin with the dual GluGLP-1 agonist ZP2929 causes improved glycemic control without body weight gain in db/db mice. Poster at ADA 71st Scientific Sessions 2011, 24 – 28 June, 2011, San Diego, California, USA. Diabetes 2011, 60, Suppl 1, A418, Poster 1527-P

[28] Daugaard, J.R., Riber, D., Larsen, K., *et al.,* The New Dual Glucagon-GLP-1 Agonist ZP2929 Improves Glycemic Control and Reduces Body Weight in Murine Models of Obesity and insulin resistance. Poster at EASD 46th Annual Meeting, Stockholm, Sweden, Sep. 20-24, 2010" Diabetologica, 2010, 53, Suppl 1, S354-S354

[29] Rendell M, Drincic A, Andukuri R. Alogliptin benzoate for the treatment of type 2 diabetes. Expert Opin Pharmacother. 2012 Mar;13(4):553-63.

[30] Defronzo RA, Burant CF, Fleck P, Wilson C, Mekki Q, Pratley RE. Efficacy and tolerability of the DPP-4 inhibitor alogliptin combined with pioglitazone, in metformin-treated patients with type 2 diabetes. J Clin Endocrinol Metab. 2012 Mar 14

[31] Nauck M, Ellis G, Fleck P, Wilson C, Mekki Q; Alogliptin Study 008 Group. Efficacy and safety of adding the dipeptidyl peptidase-4 inhibitor alogliptin to metformin therapy in patients with type 2 diabetes inadequately controlled with metformin monotherapy: a multicentre, randomised, double-blind, placebo-controlled study. Int J Clin Pract. 2009;63(1):46–55.

[32] Pratley R, Kipnes M, Fleck P, Wilson C, Mekki Q; Alogliptin Study 007 Group. Efficacy and safety of the dipeptidyl peptidase-4 inhibitor alogliptin in patients with type 2 diabetes

inadequately controlled by glyburide monotherapy. Diabetes Obes Metab. 2009;11(2):167–176.

[33]　Pratley RE, Reusch JE, Fleck PR, Wilson CA, Mekki Q; Alogliptin Study 009 Group. Efficacy and safety of the dipeptidyl peptidase-4 inhibitor alogliptin added to pioglitazone in patients with type 2 diabetes: a randomized, double-blind, placebo-controlled study.Curr Med Res Opin. 2009 Oct; 25 (10):2361-71.

[34]　Kutoh E, Ukai Y. Alogliptin as an initial therapy in patients with newly diagnosed, drug naïve type 2 diabetes: a randomized, control trial. Endocrine. 2012 Jun;41(3):435-41.

[35]　Seino Y, Fujita T, Hiroi S, Hirayama M, Kaku K. Alogliptin plus voglibose in Japanese patients with type 2 diabetes: a randomized, double-blind, placebo-controlled trial with an open-label, long-term extension. Curr Med Res Opin. 2011 Nov;27 Suppl 3:21-9.

[36]　White WB, Bakris GL, Bergenstal RM, *et al.,* Examination of cArdiovascular outcoMes with alogliptIN *versus* standard of carE in patients with type 2 diabetes mellitus and acute coronary syndrome (EXAMINE): a cardiovascular safety study of the dipeptidyl peptidase 4 inhibitor alogliptin in patients with type 2 diabetes with acute coronary syndrome. Am Heart J. 2011 Oct;162(4):620-626.e1.

[37]　Lim KS, Cho JY, Kim BH, *et al.,* Pharmacokinetics and pharmacodynamics of LC15-0444, a novel dipeptidyl peptidase IV inhibitor, after multiple dosing in healthy volunteers. Br J Clin Pharmacol. 2009 Dec;68(6):883-90.

[38]　Rhee EJ, Lee WY, Yoon KH, *et al.,* A multicenter, randomized, placebo-controlled, double-blind phase II trial evaluating the optimal dose, efficacy and safety of LC 15-0444 in patients with type 2 diabetes. Diabetes Obes Metab. 2010 Dec;12(12):1113-9.

[39]　Nair S, Wilding JP. Sodium glucose cotransporter 2 inhibitors as a new treatment for diabetes mellitus. J Clin Endocrinol Metab. 2010 Jan;95(1):34-42.

[40]　Chao EC, Henry RR. SGLT2 inhibition--a novel strategy for diabetes treatment. Nat Rev Drug Discov. 2010 Jul;9(7):551-9.

[41]　Tahrani AA, Barnett AH. Dapagliflozin: a sodium glucose cotransporter 2 inhibitor in development for type 2 diabetes. Diabetes Ther. 2010 Dec;1(2):45-56.

[42]　Komoroski B, Vachharajani N, Feng Y, Li L, Kornhauser D, Pfister M. Dapagliflozin, a novel, selective SGLT2 inhibitor, improved glycemic control over 2 weeks in patients with type 2 diabetes mellitus. Clin Pharmacol Ther. 2009 May;85(5):513-9.

[43]　Henry RR, Murray AV, Marmolejo MH, Hennicken D, Ptaszynska A, List JF. Dapagliflozin, metformin XR, or both: initial pharmacotherapy for type 2 diabetes, a randomised controlled trial. Int J Clin Pract. 2012 May;66(5):446-56.

[44]　Ferrannini E, Ramos SJ, Salsali A, Tang W, List JF. Dapagliflozin monotherapy in type 2 diabetic patients with inadequate glycemic control by diet and exercise: a randomized, double-blind, placebo-controlled, phase 3 trial. Diabetes Care. 2010 Oct;33(10):2217-24.

[45]　Bailey CJ, Gross JL, Pieters A, Bastien A, List JF. Effect of dapagliflozin in patients with type 2 diabetes who have inadequate glycaemic control with metformin: a randomised, double-blind, placebo-controlled trial. Lancet. 2010 Jun 26;375(9733):2223-33.

[46]　Wilding JP, Woo V, Soler NG, *et al.,* Dapagliflozin 006 Study Group. Long-term efficacy of dapagliflozin in patients with type 2 diabetes mellitus receiving high doses of insulin: a randomized trial.

[47]　Strojek K, Yoon KH, Hruba V, Elze M, Langkilde AM, Parikh S. Effect of dapagliflozin in patients with type 2 diabetes who have inadequate glycaemic control with glimepiride: a randomized, 24-week, double-blind, placebo-controlled trial. Diabetes Obes Metab. 2011 Oct;13(10):928-38.

[48] Nauck MA, Del Prato S, Meier JJ, *et al.,* Dapagliflozin *versus* glipizide as add-on therapy in patients with type 2 diabetes who have inadequate glycemic control with metformin: a randomized, 52-week, double-blind, active-controlled noninferiority trial. Diabetes Care. 2011 Sep;34(9):2015-22

[49] Rosenstock J, Vico M, Wei L, Salsali A, List JF. Effects of dapagliflozin, a sodium-glucose cotransporter-2 inhibitor, on hemoglobin a1c, body weight, and hypoglycemia risk in patients with type 2 diabetes inadequately controlled on pioglitazone monotherapy. Diabetes Care. 2012 Mar 23.

[50] Liang Y, Arakawa K, Ueta K, *et al.,* Effect of canagliflozin on renal threshold for glucose, glycemia, and body weight in normal and diabetic animal models. PLoS One. 2012;7(2):e30555.

[51] Devineni D, Morrow L, Hompesch M, *et al.,* Canagliflozin improves glycaemic control over 28 days in subjects with type 2 diabetes not optimally controlled on insulin. Diabetes Obes Metab. 2012 Jun;14(6):539-45.

[52] Rosenstock J, Aggarwal N, Polidori D, *et al.,* Dose-Ranging Effects of Canagliflozin, a Sodium-Glucose Cotransporter 2 Inhibitor, as Add-On to Metformin in Subjects With Type 2 Diabetes. Diabetes Care. 2012 Apr 9.

[53] Imamura M, Nakanishi K, Suzuki T, *et al.,* Discovery of Ipragliflozin (ASP1941): A novel C-glucoside with benzothiophene structure as a potent and selective sodium glucose co-transporter 2 (SGLT2) inhibitor for the treatment of type 2 diabetes mellitus. Bioorg Med Chem. 2012 May 15;20(10):3263-79.

[54] Veltkamp SA, Kadokura T, Krauwinkel WJ, Smulders RA. Effect of Ipragliflozin (ASP1941), a novel selective sodium-dependent glucose co-transporter 2 inhibitor, on urinary glucose excretion in healthy subjects. Clin Drug Investig. 2011 Dec 1;31(12):839-51.

[55] Grempler R, Thomas L, Eckhardt M, *et al.,* Empagliflozin, a novel selective sodium glucose cotransporter-2 (SGLT-2) inhibitor: characterisation and comparison with other SGLT-2 inhibitors. Diabetes Obes Metab. 2012 Jan;14(1):83-90.

[56] R.M. Jones, J.N. Leonard, D.J. Buzard, J. Lehmann, GPR119 agonists for the treatment of type 2 diabetes, Expert Opin. Ther. Pat. 19 (2009) 1339–1359.

[57] Overton HA, Fyfe MC and Reynet C. GPR119, a novel G protein-coupled receptor target for the treatment of type 2 diabetes and obesity. Br J Pharmacol. 2008 March; 153(S1): S76–S81.

[58] Yoshida S, Ohishi T, Matsui T, *et al.,* The role of small molecule GPR119 agonist, AS1535907, in glucose-stimulated insulin secretion and pancreatic β-cell function. Diabetes Obes Metab. 2011 Jan;13(1):34-41

[59] Katz LB, Gambale JJ, Rothenberg PL, *et al.,* Effects of JNJ-38431055, a novel GPR119 receptor agonist, in randomized, double-blind, placebo-controlled studies in subjects with type 2 diabetes. Diabetes Obes Metab. 2012 Feb 16.

[60] Gribble FM. Alpha2A-adrenergic receptors and type 2 diabetes. N Engl J Med. 2010 Jan 28;362(4):361-2.

[61] Rosengren AH, Jokubka R, Tojjar D, *et al.,* Overexpression of alpha2A-adrenergic receptors contributes to type 2 diabetes. Science. 2010 Jan 8;327(5962):217-20.

[62] Abdel-Zaher AO, Ahmed IT, El-Koussi AD. The potential antidiabetic activity of some alpha-2 adrenoceptor antagonists. Pharmacol Res. 2001 Nov;44(5):397-409.

[63] Nam DH, Lee MH, Kim JE, *et al.,* Blockade of cannabinoid receptor 1 improves insulin resistance, lipid metabolism, and diabetic nephropathy in db/db mice. Endocrinology. 2012 Mar;153(3):1387-96.

[64]   Scheen AJ. CB1 receptor blockade and its impact on cardiometabolic risk factors: overview of the RIO programme with rimonabant. J Neuroendocrinol. 2008 May;20 Suppl 1:139-46.

[65]   Kim W, Doyle ME, Liu Z, *et al.,* Cannabinoids inhibit insulin receptor signaling in pancreatic β-cells. Diabetes. 2011 Apr;60(4):1198-209.

[66]   Horváth B, Mukhopadhyay P, Haskó G, Pacher P. The endocannabinoid system and plant-derived cannabinoids in diabetes and diabetic complications. Am J Pathol. 2012 Feb;180(2):432-42.

[67]   Matschinsky FM, Zelent B, Doliba N, *et al.,* Glucokinase activators for diabetes therapy: May 2010 status report. Diabetes Care. 2011 May;34 Suppl 2:S236-43

[68]   Matschinsky FM, Zelent B, Doliba NM, *et al.,* Research and development of glucokinase activators for diabetes therapy: theoretical and practical aspects. Handb Exp Pharmacol. 2011;(203):357-401.

[69]   Liu S, Ammirati MJ, Song X, *et al.,* Insights into the mechanism of glucokinase activation: observation of multiple distinct protein conformations. J Biol Chem. 2012 Feb 1.

[70]   Pal M. Recent advances in glucokinase activators for the treatment of type 2 diabetes. Drug Discov Today. 2009 Aug;14(15-16):784-92.

[71]   Pike KG, Allen JV, Caulkett PW, *et al.,* Design of a potent, soluble glucokinase activator with increased pharmacokinetic half-life. Bioorg Med Chem Lett. 2011 Jun 1;21(11):3467-70.

[72]   Johnson TO, Ermolieff J, Jirousek MR. Protein tyrosine phosphatase 1B inhibitors for diabetes. Nat Rev Drug Discov. 2002 Sep;1(9):696-709.

[73]   Zhang ZY, Lee SY.PTP1B inhibitors as potential therapeutics in the treatment of type 2 diabetes and obesity. Expert Opin Investig Drugs. 2003 Feb;12(2):223-33.

[74]   Koren S, Fantus IG. Inhibition of the protein tyrosine phosphatase PTP1B: potential therapy for obesity, insulin resistance and type-2 diabetes mellitus. Best Pract Res Clin Endocrinol Metab. 2007 Dec;21(4):621-40

[75]   Ma YM, Tao RY, Liu Q, *et al.,* PTP1B inhibitor improves both insulin resistance and lipid abnormalities *in vivo* and *in vitro*. Mol Cell Biochem. 2011 Nov;357(1-2):65-72. Epub 2011 May 21

[76]   Jiang B, Shi D, Cui Y, Guo S. Design, Synthesis, and Biological Evaluation of Bromophenol Derivatives as Protein Tyrosine Phosphatase 1B Inhibitors. Arch Pharm (Weinheim). 2012 Feb 3.

[77]   Xie J, Tian J, Su L, Huang M, Zhu X, Ye F, Wan Y. Pyrrolo[2,3-c]azepine derivatives: a new class of potent protein tyrosine phosphatase 1B inhibitors. Bioorg Med Chem Lett. 2011 Jul 15;21(14):4306-9.

[78]   Patel D, Jain M, Shah SR, *et al.,* Discovery of orally active, potent, and selective benzotriazole-based PTP1B inhibitors. ChemMedChem. 2011 Jun 6;6(6):1011-6.

[79]   Forghieri M, Laggner C, Paoli P, *et al.,* Synthesis, activity and molecular modeling of a new series of chromones as low molecular weight protein tyrosine phosphatase inhibitors. Bioorg Med Chem. 2009 Apr 1;17(7):2658-72.

[80]   Mandrup-Poulsen T. The role of interleukin-1 in the pathogenesis of IDDM. Diabetologia 1996;39:1005-29.

[81]   Welsh N, Cnop M, Kharroubi I, *et al.* Is there a role for locally produced interleukin-1 in the deleterious effects of high glucose or the type 2 diabetes milieu to human pancreatic islets? Diabetes 2005;54: 3238-44.

[82]   Dinarello CA. The role of the interleukin-1–receptor antagonist in blocking inflammation mediated by interleukin-1. N Engl J Med 2000;343:732-4.

[83]   Zumsteg U, Reimers II, Pociot F, *et al.* Differential interleukin-1 receptor antagonism on pancreatic beta and alpha cells:studies in rodent and human islets and innormal rats. Diabetologia 1993;36:759-66.

[84]   Maedler K, Sergeev P, Ris F, *et al.,* Glucose-induced β-cell production of IL-1β contributes to glucotoxicity in human pancreatic islets. J Clin Invest 2002;110:851–860

[85]   Larsen CM, Faulenbach M, Vaag A, *et al.,* Interleukin-1-receptor antagonist in type 2 diabetes mellitus. N Engl J Med. 2007 Apr 12;356(15):1517-26.

[86]   Larsen CM, Faulenbach M, Vaag A, Ehses JA, Donath MY, Mandrup-Poulsen T. Sustained effects of interleukin-1 receptor antagonist treatment in type 2 diabetes. Diabetes Care. 2009 Sep; 32(9):1663-8.

[87]   Spranger J, Kroke A, Möhlig M, *et al.,* Inflammatory cytokines and the risk to develop type 2 diabetes: results of the prospective population-based European Prospective Investigation into Cancer and Nutrition (EPIC)-Potsdam Study. Diabetes. 2003 Mar;52(3):812-7.

[88]   Vincent JA, Mohr S. Inhibition of caspase-1/interleukin-1beta signaling prevents degeneration of retinal capillaries in diabetes and galactosemia. Diabetes. 2007 Jan;56(1):224-30.

[89]   Abbate A, Salloum FN, Vecile E, *et al.,* Anakinra, a recombinant human interleukin-1 receptor antagonist, inhibits apoptosis in experimental acute myocardial infarction. Circulation. 2008 May 20;117(20):2670-83.

[90]   van Poelje PD, Dang Q, Erion MD. Discovery of fructose-1,6-bisphosphatase inhibitors for the treatment of type 2 diabetes. Curr Opin Drug Discov Devel. 2007 Jul;10(4):430-7

[91]   Zhang Y, Xie Z, Zhou G, Zhang H, Lu J, Zhang WJ. Fructose-1,6-bisphosphatase regulates glucose-stimulated insulin secretion of mouse pancreatic beta-cells. Endocrinology. 2010 Oct;151(10):4688-95.

[92]   Dang Q, Kasibhatla SR, Reddy KR, *et al.,* Discovery of potent and specific fructose-1,6-bisphosphatase inhibitors and a series of orally-bioavailable phosphoramidase-sensitive prodrugs for the treatment of type 2 diabetes. J Am Chem Soc. 2007 Dec 19; 129(50):15491-502.

[93]   Erion MD, van Poelje PD, Dang Q, *et al.,* MB06322 (CS-917): A potent and selective inhibitor of fructose 1,6-bisphosphatase for controlling gluconeogenesis in type 2 diabetes. Proc Natl Acad Sci U S A. 2005 May 31; 102(22):7970-5.

[94]   van Poelje PD, Potter SC, Chandramouli VC, Landau BR, Dang Q, Erion MD. Inhibition of fructose 1,6-bisphosphatase reduces excessive endogenous glucose production and attenuates hyperglycemia in Zucker diabetic fatty rats. Diabetes. 2006 Jun;55(6):1747-54.

[95]   Yoshida T, Okuno A, Izumi M, *et al.,* CS-917, a fructose 1,6-bisphosphatase inhibitor, improves postprandial hyperglycemia after meal loading in non-obese type 2 diabetic Goto-Kakizaki rats. Eur J Pharmacol. 2008 Dec 28;601(1-3):192-7.

[96]   Yoshida T, Okuno A, Takahashi K, *et al.,* Contributions of hepatic gluconeogenesis suppression and compensative glycogenolysis on the glucose-lowering effect of CS-917, a fructose 1,6-bisphosphatase inhibitor, in non-obese type 2 diabetes Goto-Kakizaki rats. J Pharmacol Sci. 2011;115(3):329-35.

[97]   van Poelje PD, Potter SC, Erion MD. Fructose-1, 6-bisphosphatase inhibitors for reducing excessive endogenous glucose production in type 2 diabetes. Handb Exp Pharmacol. 2011;(203):279-301.

[98]   Fouqueray P, Leverve X, Fontaine E, *et al.* (2011) Imeglimin - A New Oral Anti-Diabetic that Targets the Three Key Defects of type 2 Diabetes. J Diabetes Metab 2:126.

[99]   Pirags V, Lebovitz H and Fouqueray P. Imeglimin, a novel glimin oral anti-diabetic, exhibits a good efficacy and safety profile in type 2 diabetic patients. Diabetes Obes Metab. 2012 Apr

[100]   Pirags V, Lebovitz H, Fouqueray P (2010) Imeglimin, a novel glimin oral antidiabetic, exhibits good glycaemic control in type 2 diabetes mellitus patients. Presented at the 46th EASD Annual Meeting, Stockholm, Sweden.

[101]   Zhang BB, Zhou G, Li C. AMPK: an emerging drug target for diabetes and the metabolic syndrome. Cell Metab. 2009 May;9(5):407-16

[102]   Misra P, Chakrabarti R. The role of AMP kinase in diabetes. Indian J Med Res. 2007 Mar;125(3):389-98.

[103]   Benoît Viollet, Louise Lantier, Jocelyne Devin-Leclerc, *et al.,* Targeting the AMPK pathway for the treatment of Type 2 diabetes. Front Biosci. 2009; 14: 3380–3400.

[104]   Nguyen PH, Gauhar R, Hwang SL, *et al.,* New dammarane-type glucosides as potential activators of AMP-activated protein kinase (AMPK) from Gynostemma pentaphyllum. Bioorg Med Chem. 2011 Nov 1;19(21):6254-60.

[105]   Hardie DG. Sensing of energy and nutrients by AMP-activated protein kinase. Am J Clin Nutr. 2011 Apr;93(4):891S-6.

[106]   Sanders MJ, Ali ZS, Hegarty BD, Heath R, Snowden MA, Carling D. Defining the mechanism of activation of AMP-activated protein kinase by the small molecule A-769662, a member of the thienopyridone family. J Biol Chem. 2007 Nov 9;282(45):32539-48.

[107]   Kershaw EE, Morton NM, Dhillon H, Ramage L, Seckl JR, Flier JS. Adipocyte-specific glucocorticoid inactivation protects against diet-induced obesity. Diabetes 2005;54:1023–103

[108]   Morton NM, Paterson JM, Masuzaki H, *et al.,* Novel adipose tissue-mediated resistance to diet-induced visceral obesity in 11-hydroxysteroid dehydrogenase type 1–deficient mice. Diabetes 2004;53:931–938

[109]   Ge R, Huang Y, Liang G, Li X. 11beta-hydroxysteroid dehydrogenase type 1 inhibitors as promising therapeutic drugs for diabetes: status and development. Curr Med Chem. 2010;17(5):412-22.

[110]   Morgan SA, Tomlinson JW. 11β-hydroxysteroid dehydrogenase type 1 inhibitors for the treatment of type 2 diabetes. Expert Opin Investig Drugs. 2010 Sep; 19(9): 1067-76.

[111]   Tiwari A. INCB-13739, an 11beta-hydroxysteroid dehydrogenase type 1 inhibitor for the treatment of type 2 diabetes. IDrugs. 2010 Apr;13(4):266-75

[112]   Hollis G, Huber R.11β-Hydroxysteroid dehydrogenase type 1 inhibition in type 2 diabetes mellitus. Diabetes Obes Metab. 2011 Jan;13(1):1-6.

[113]   Rosenstock J, Banarer S, Fonseca VA, *et al.,* The 11-beta-hydroxysteroid dehydrogenase type 1 inhibitor INCB13739 improves hyperglycemia in patients with type 2 diabetes inadequately controlled by metformin monotherapy. Diabetes Care. 2010 Jul;33(7):1516-22.

[114]   Feig PU, Shah S, Hermanowski-Vosatka A, *et al.,* Effects of an 11β-hydroxysteroid dehydrogenase type 1 inhibitor, MK-0916, in patients with type 2 diabetes mellitus and metabolic syndrome. Diabetes Obes Metab. 2011 Jun; 13(6):498-504.

[115]   Shah S, Hermanowski-Vosatka A, Gibson K, *et al.,* Efficacy and safety of the selective 11β-HSD-1 inhibitors MK-0736 and MK-0916 in overweight and obese patients with hypertension. J Am Soc Hypertens. 2011 May-Jun;5(3):166-76.

[116] Gibbs JP, Emery MG, McCaffery I, *et al.,* Population pharmacokinetic/pharmacodynamic model of subcutaneous adipose 11β-hydroxysteroid dehydrogenase type 1 (11β-HSD1) activity after oral administration of AMG 221, a selective 11β-HSD1 inhibitor. J Clin Pharmacol. 2011 Jun;51(6):830-41.

[117] Adeghate E, Adem A, Hasan MY, Tekes K, Kalasz H. Medicinal Chemistry and Actions of Dual and Pan PPAR Modulators. Open Med Chem J. 2011;5(Suppl 2):93-8.

[118] Chang F, Jaber LA, Berlie HD, O'Connell MB. Evolution of peroxisome proliferator-activated receptor agonists. Ann Pharmacother. 2007 Jun;41(6):973-83.

[119] Cavender MA, Lincoff AM. Therapeutic potential of aleglitazar, a new dual PPAR-α/γ agonist: implications for cardiovascular disease in patients with diabetes mellitus. Am J Cardiovasc Drugs. 2010;10(4):209-16.

[120] Henry RR, Lincoff AM, Mudaliar S, Rabbia M, Chognot C, Herz M. Effect of the dual peroxisome proliferator-activated receptor-alpha/gamma agonist aleglitazar on risk of cardiovascular disease in patients with type 2 diabetes (SYNCHRONY): a phase II, randomised, dose-ranging study. Lancet. 2009 Jul 11;374(9684):126-35.

[121] Hansen BC, Tigno XT, Bénardeau A, Meyer M, Sebokova E, Mizrahi J. Effects of aleglitazar, a balanced dual peroxisome proliferator-activated receptor α/γ agonist on glycemic and lipid parameters in a primate model of the metabolic syndrome. Cardiovasc Diabetol. 2011 Jan 20;10:7.

[122] Carolina Pellanda, Dennis Ruff, Ruth Penn, *et al.,* Concomitant treatment with aleglitazar does not change the effects on renal function induced by aspirin. J Am Coll Cardiol, 2012; 59:1703

[123] Li PP, Shan S, Chen YT, *et al.,* The PPARalpha/gamma dual agonist chiglitazar improves insulin resistance and dyslipidemia in MSG obese rats. Br J Pharmacol. 2006 Jul;148(5):610-8.

[124] Jeong HW, Lee JW, Kim WS, *et al.,* A nonthiazolidinedione peroxisome proliferator-activated receptor α/γ dual agonist CG301360 alleviates insulin resistance and lipid dysregulation in db/db mice. Mol Pharmacol. 2010 Nov;78(5):877-85.

[125] Heal DJ, Gosden J, Smith SL. What is the prognosis for new centrally-acting anti-obesity drugs? Neuropharmacology. 2012 Jul;63(1):132-46.

[126] Bays HE, Gadde KM. Phentermine/topiramate for weight reduction and treatment of adverse metabolic consequences in obesity. Drugs Today (Barc). 2011 Dec;47(12):903-14. Review.

[127] Gadde KM, Allison DB, Ryan DH, *et al.,* Effects of low-dose, controlled-release, phentermine plus topiramate combination on weight and associated comorbidities in overweight and obese adults (CONQUER): a randomised, placebo-controlled, phase 3 trial. Lancet. 2011 Apr 16;377(9774):1341-52.

[128] Garvey WT, Ryan DH, Look M, *et al.,* Two-year sustained weight loss and metabolic benefits with controlled-release phentermine/topiramate in obese and overweight adults (SEQUEL): a randomized, placebo-controlled, phase 3 extension study. Am J Clin Nutr. 2012 Feb;95(2):297-308.

[129] Powell AG, Apovian CM, Aronne LJ. The combination of phentermine and topiramate is an effective adjunct to diet and lifestyle modification for weight loss and measures of comorbidity in overweight or obese adults with additional metabolic risk factors. Evid Based Med. 2012 Feb;17(1):14-5.

[130] Martin CK, Redman LM, Zhang J, *et al.,* Lorcaserin, a 5-HT(2C) receptor agonist, reduces body weight by decreasing energy intake without influencing energy expenditure. J Clin Endocrinol Metab. 2011 Mar;96(3):837-45.

[131] Bai B, Wang Y. The use of lorcaserin in the management of obesity: a critical appraisal. Drug Des Devel Ther. 2010 Dec 20;5:1-7.

[132] O'Neil PM, Smith SR, Weissman NJ, *et al.,* Randomized Placebo-Controlled Clinical Trial of Lorcaserin for Weight Loss in Type 2 Diabetes Mellitus: The BLOOM-DM Study. Obesity (Silver Spring). 2012 Jul;20(7):1426-36.

[133] Weir GC, Cavelti-Weder C, Bonner-Weir S. Stem cell approaches for diabetes: towards beta cell replacement. Genome Med. 2011 Sep 27;3(9):61.

[134] Wang L, Zhao S, Mao H, Zhou L, Wang ZJ, Wang HX. Autologous bone marrow stem cell transplantation for the treatment of type 2 diabetes mellitus. Chin Med J (Engl). 2011 Nov;124(22):3622-8.

[135] Bhansali A, Upreti V, Khandelwal N, *et al.,* Efficacy of autologous bone marrow-derived stem cell transplantation in patients with type 2 diabetes mellitus. Stem Cells Dev. 2009 Dec;18(10):1407-16.

[136] Jiang R, Han Z, Zhuo G, *et al.,* Transplantation of placenta-derived mesenchymal stem cells in type 2 diabetes: a pilot study. Front Med. 2011 Mar;5(1):94-100.

[138] Elder KA, Wolfe BM. Bariatric surgery: a review of procedures and outcomes. Gastroenterology. 2007 May;132(6):2253-71.

[138] Laferrère B. Diabetes remission after bariatric surgery: is it just the incretins? Int J Obes (Lond). 2011 Sep;35 Suppl 3:S22-5.

[139] Buchwald H, Estok R, Fahrbach K, *et al.,* Weight and type 2 diabetes after bariatric surgery: systematic review and meta-analysis. Am J Med 2009;122:248–256

[140] Meijer RI, van Wagensveld BA, Siegert CE, Eringa EC, Serné EH, Smulders YM. Bariatric surgery as a novel treatment for type 2 diabetes mellitus: a systematic review. Arch Surg. 2011 Jun;146(6):744-50.

[141] Dixon JB, le Roux CW, Rubino F, Zimmet P. Bariatric surgery for type 2 diabetes. Lancet. 2012 Jun 16;379(9833):2300-11.

[142] Dixon JB, Zimmet P, Alberti KG, Rubino F; International Diabetes Federation Taskforce on Epidemiology and Prevention. Bariatric surgery: an IDF statement for obese Type 2 diabetes. Diabet Med. 2011 Jun;28(6):628-42.

[143] Shimizu H, Timratana P, Schauer PR, Rogula T. Review of Metabolic Surgery for Type 2 Diabetes in Patients with a BMI < 35 kg/m(2). J Obes. 2012;2012:147256.

[144] Strader AD. Ileal transposition provides insight into the effectiveness of gastric bypass surgery. Physiol Behav. 2006 Jun 30;88(3):277-82.

[145] Wang TT, Hu SY, Gao HD, *et al.,* Ileal transposition controls diabetes as well as modified duodenal jejunal bypass with better lipid lowering in a nonobese rat model of type II diabetes by increasing GLP-1. Ann Surg. 2008 Jun;247(6):968-75.

[146] DePaula AL, Stival AR, DePaula CC, Halpern A, Vencio S. Surgical treatment of type 2 diabetes in patients with BMI below 35: mid-term outcomes of the laparoscopic ileal interposition associated with a sleeve gastrectomy in 202 consecutive cases. J Gastrointest Surg. 2012 May;16(5):967-76.

[147] Klein S, Fabbrini E, Patterson BW, *et al.,* Moderate effect of duodenal-jejunal bypass surgery on glucose homeostasis in patients with type 2 diabetes. Obesity (Silver Spring). 2012 Jun;20(6):1266-72.

CHAPTER 6

Advances in New Drugs for Diabetes Treatment: GLP-1 Mimetics and DPP-4 Inhibitors

Xianglan Sun[1], Yao Wang[2] and Minglong Li[1,*]

[1]Department of Endocrinology, Provincial Hospital affiliated to Shandong University, Jinan, China and [2]Shandong Institute of Endocrine and Metabolic Diseases, Shandong Academy of Medical Sciences, Jinan, 250062 Shandong, China

Abstract: The dramatic rise in the prevalence of obesity and diabetes is associated with increased mortality, morbidity as well as public health care costs worldwide. The need for new effective and long-lasting drugs is urgent. Recent research has focused on the role of incretin in the maintenance of glucose homeostasis through their actions on both α- and β-cell function. Moreover, increased knowledge of the pathophysiology of diabetes has contributed to the development of novel drugs, including injectable glucagon-like peptide-1 (GLP-1) mimetics, and oral dipeptidyl peptidase-4 (DPP-4) inhibitors. GLP-1 agonists mimic the effect of this incretin, whereas DPP-4 inhibitors prevent the inactivation of the endogenously released hormone. GLP-1 appears to be involved in both peripheral and central pathways mediating satiety. Clinical trials have shown that two GLP-1 receptor agonists exenatide and liraglutide have a weight-lowering potential in non-diabetic obese individuals. Furthermore, they may also hold a potential in preventing diabetes as compared to other weight loss agents. Both agents offer an effective alternative to the currently available hypoglycaemic drugs but further evaluation is needed to confirm their clinical roles and safety. The purpose of this review is to summarize the background of the incretin, and the roles and side effects of the GLP-1 agonists and DPP-4 inhibitors in clinical trials. In addition, up-to-date literature on GLP-1 agonists and DPP-4 inhibitors based clinical therapies will be summarized with a special mention of their weight-lowering properties. In conclusion, the incretin impairment, which seems to exist in both obesity and diabetes, may link these two pathologies and underlines the potential of incretin based therapies in the prevention and treatment of obesity and diabetes.

Keywords: Alogliptin, DPP-4 Inhibitors, Exenatide, GLP-1 Agonists, Incretin, Linagliptin, Liraglutide, Saxagliptin, Sitagliptin, Vildagliptin.

INTRODUCTION

The prevalence of obesity and type 2 diabetes mellitus (T2DM) are evolving globally at an alarming rate in the 21st century as a result of overnutrition, an

***Corresponding author Minglong Li:** Department of Endocrinology, Provincial Hospital affiliated to Shandong University, Jinan, China; Tel:+86 531 68777160; Fax:+86 531 68777160; E-mail: ellinasun98@sina.com*

Atta-ur-Rahman (Ed)

ageing population, lack of exercise, and increased migration of susceptible patients. The International Diabetes Federation estimated in 2025 that 380 million adults worldwide had diabetes [1]. However, the global statistics of diabetes mellitus in year 2013 has indicated already, about 382 million people had this disease worldwide [2]. In year 2012 and 2013 diabetes resulted in mortality of 1.5-5.1 million people per year, making it the 8th leading cause of death in the world [2]. And the World Health Organization (WHO) has estimated the prevalence of DM in the USA in 2025 is 11.2% [3]. In addition, an average age of onset of diabetes is about 42.5 years and the economic cost of diabetes seems to have a significant increased worldwide. The primary aim of managing T2DM is to delay, or even prevent the complications of the disease by achieving good glycaemic control. Therefore, the need for new effective and long-lasting anti-diabetic drugs is urgent.

Increased knowledge of pathophysiology of T2DM has contributed to the development of novel treatments, such as glucagon-like peptide-1 (GLP-1) mimetics and dipeptidyl peptidase-4 (DPP-4) inhibitors. GLP-1 agonists mimic the effect of incretin, whereas DPP-4 inhibitors prevent the inactivation of the endogenously released hormone. Both agents offer an effective alternative to the currently available hypoglycaemic drugs. And they may also hold a potential role in preventing obesity and diabetes as compared to other weight loss agents. However, further evaluation is needed to confirm their clinical roles and safety. The aim of this review is to summarize the background of the incretin, and explore the roles and side effects of the GLP-1 agonists and DPP-4 inhibitors. In addition, their weight-lowering properties will be summarized specially.

INCRETINS

Incretins are a group of metabolic hormones that stimulate a decrease in blood glucose levels (in Fig. **1**). And they are gut-derived peptides secreted in response to meals. The major incretins are GLP-1 and glucose dependent insulinotropic peptide (GIP) and they account for approximately 90% of the incretin activity [4]. Both GLP-1 and GIP are rapidly inactivated by the enzyme DPP-4. The GLP-1is produced by L-cells in the distal gut, and the GIP is produced by duodenal K-cells in response to ingested nutrients [5]. The incretins are secreted into the circulation within minutes in response to a meal and up on release they bind to specific G-protein coupled receptors present on cells and other target tissues [6]. GLP-1 is secreted in greater concentrations than GIP and is considered more physiologically relevant in humans [7]. In β-cells, GLP-1 enhances glucose-dependent insulin secretion, increases insulin synthesis, and in animals stimulates β-cell proliferation and inhibits apoptosis [5]. GLP-1 also reduces glucose concentrations through inhibition of pancreatic α-cell glucagon secretion and indirectly via inhibition of gastric emptying and appetite [8, 9]. In addition to slowing gastric emptying, GLP-1 may also decrease intestinal lymph flow, triglyceride absorption, and apolipoprotein synthesis adding to a complex combination of mechanisms that may limit the release of triglycerides into the circulation after lipid-containing meals [10].

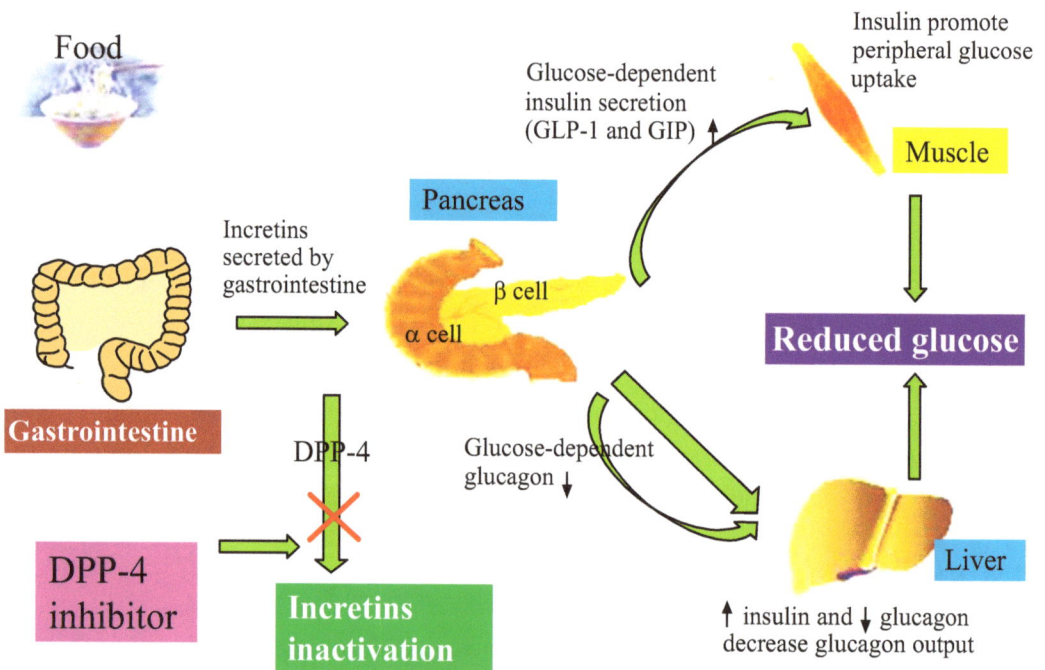

Figure 1: The effects of incretins in pathophysiology of T2DM.

GLP-1 levels are significantly decreased in T2DM (approximately 50% compared to healthy individuals) [11, 12]. Therefore, the incretin-based therapy has emerged as a strategy in managing T2DM, primarily because they generally do not cause hypoglycemia and possess weight-neutral or weight losing properties. And the efficacy of these agents is more or less similar to commonly used drugs metformin and sulfonylureas. Early GLP-1 therapy was suggested to preserve β - cell function in subjects with impaired glucose tolerance (IGT) or mild T2DM [13]. Interestingly, a recent meta-analysis suggested that Asians may have better response to these incretin-based therapies [14].

THE ROLES OF THE GLP-1 AGONISTS

GLP-1 is a 30-amino acid peptide produced in the intestinal epithelial L-cells of the distal ileum and colon by differential processing of the proglucagon gene from the prohormone convertase PC1/3 [15]. GLP-1 binds to GLP-1 receptor in the cell membrane of the pancreatic islets [16]. And it exerts its action on β-cell proliferation and survival by phosphatidylinositol-3 kinase (PI3-K) and protein kinase B (PKB/Akt), p38 mitogen-activated protein kinase (MAPK) and protein kinase Cζ pathways [17, 18].

GLP-1 amide is not very useful for treatment of type 2 diabetes mellitus, since it must be administered by continuous subcutaneous infusion. Therefore, several long-lasting analogs have been developed. So far, exenatide and liraglutide have been approved for use widely. These agents have benefits of a lower risk of hypoglycemia, and having a potential for weight reduction. Moreover, these incretin- based therapies also exert anti-inflammatory activity [19]. However, the main disadvantage of these GLP-1 analogs is they must be administered by subcutaneous injection. Exenatide is administered twice daily or once weekly as a microsphere preparation, and liraglutide is a once-daily formulation.

Exenatide

Exenatide, the first FDA approved GLP-1mimetic,is the synthetic form of the naturally occurring exendin-4, a 39-amino-acid peptide hormone secreted by the salivary glands of the venomous lizard Heloderma suspectum [20]. It has only 53% homology to the human GLP-1 amino acid sequence. So it is relatively more resistant to DPP-4, reaching a maximum level approximately 2 hours following subcutaneous injection [21]. Exenatide plays many of the actions of GLP-1, such as enhancement of glucose induced insulin secretion, inhibition of glucagon release, reduction of fasting and postprandial glucose, delay of gastric emptying, inhibition of appetite and induction of weight loss [21-24]. Moreover, exenatide was associated with significant improvement in multiple cardiovascular risk factors including systolic and diastolic blood pressure, fasting triglycerides, as well as total, LDL- and HDL-cholesterol [25]. In addition, the anti-atherosclerotic effect of exenatide was associating with inhibition of inflammatory responses of atherosclerotic plaque macrophages [26]. In a word, exenatide played important roles in obese and T2DM in a large number of clinical trials.

One hundred fifty two obese (BMI: 39.6 ± 7.0 kg/m^2) individuals were randomized to receive either exenatide or placebo, along with lifestyle modification for 24 weeks [27]. Exenatide treated individuals lost 5.1 ± 0.5 kg from baseline *vs.* 1.6 ± 0.5 kg in the placebo group. An important percentage of individuals with prediabetes returned to normal glucose after the end of the period. Therefore, exenatide therapy in addition to lifestyle modification is a promising therapeutic approach for obese prediabetic individuals. In another non randomized study, 105 individuals with IGT and/or IFG were treated with: (1) Lifestyle modification only ;(2) Pioglitazone 15mg daily and metformin 850mg daily; and (3) A triple combination of pioglitazone, metformin and exenatide [28]. In the pioglitazone and metformin group, insulin sensitivity and β-cell function improved by 42% and 50% respectively, while 14% of the individuals with IGT and 36% of the individuals with IFG reverted to NGT. Interestingly, in the triple therapy group, a robust 109% improvement in β-cell function and a 52% increased in insulin sensitivity was observed, while 59% of the individuals with IGT and 56% of the individuals with IFG reverted to NGT. In addition, a 24-week prospective randomized outpatient clinical trial explored the possible role of exenatide and metformin, alone or in combination, in 60 overweight/obese women with polycystic ovary syndrome [29]. At the end of the study, the combination arm experienced weight loss of 6 ± 0.5 kg, the exenatide arm 3.2 ± 0.1 kg, and the metformin arm 1.6 ± 0.2 kg. And the insulin secretion was significantly reduced in the exenatide and combination arms.

Exenatide improved postprandial endothelial function and reduced postprandial rises in insulin, glucose and triglycerides concentrations in individuals with IGT and patients with recent T2DM in a double-blinded randomized crossover study [30]. Exenatide also reduced the postprandial elevation of triglycerides, apolipoprotein B-48, remnant lipoprotein cholesterol and remnant lipoprotein triglyceride in individuals with IGT and patients with recent onset T2DM [31]. Both studies suggested an additional cardiovascular benefit of this agent beyond the improved glycemic control in this population. Another randomized 3-week head-to-head study examined the effects of exenatide and metformin on microvascular endothelial function in 50 individuals with abdominal obesity and prediabetes [32]. Similar effects of both agents were shown on microvascular endothelial function, vascular activation, oxidative stress and markers inflammation.

Exenatide had greater HbA$_{1c}$ reductions compared with optimized insulin glargine alone, irrespective of baseline HbA$_{1c}$ [33]. Exenatide participants lost more weight, regard less of baseline HbA$_{1c}$ or BMI. Changes were evident in modestly obese patients and in those with longer diabetes duration. Moreover, a total of 60 newly diagnosed patients with obesity, non-alcoholic fatty liver disease (NAFLD) with elevated liver enzymes and T2DM were included in the study [34]. After 12 weeks, body weight and waist circumference were significantly decreased in the exenatide group but increased in the intensive insulin group. The levels of alanine aminotransferase (ALT), aspartate aminotransferase (AST) and γ-glutamyl transpeptidase (γGGT) in the exenatide group were significantly lower than in the intensive insulin group. Moreover, the reversal rate of fatty liver was significantly higher in the exenatide group (93.3%) than the intensive insulin group (66.7%). Therefore, exenatide has a better hepatic protective effect than intensive insulin therapy and perhaps represents a unique option for adjunctive therapy for patients with obesity, nonalcoholic fatty liver disease with elevated liver enzymes and T2DM.

Fifty-eight patients with ST-segment elevation myocardial infarction (STEMI) and thrombolysis in myocardial infarction flow 0 were enrolled in the study and randomly assigned to receive either exenatide or placebo (saline) subcutaneously [35]. The releases of creatine kinase-MB and troponin I were significantly reduced in the exenatide group. In 58 patients evaluated with cardiac magnetic resonance (CMR), the absolute mass of delayed hyper enhancement was significantly reduced in the exenatide group compared to the control group. Therefore, the adjunctive exenatide therapy with primary percutaneous coronary intervention was associated with reduction of infarct size and improvement of subclinical left ventricular function. A total of 172 patients with STEMI and thrombolysis in myocardial infarction flow 0/1 were randomly assigned to exenatide or placebo (saline) intravenously [36]. In 105 patients evaluated with CMR, a significantly larger salvage index was found in the exenatide group than in the placebo group. Infarct size was also smaller in the exenatide group. There was a trend towards smaller absolute infarct size in the exenatide group. No difference was observed in left ventricular function or 30-day clinical events. Therefore, in patients with STEMI, administration of exenatide at the time of reperfusion increases myocardial salvage, reduces final infarct size and short-duration of ischemia and preserve cardiac function [37].

Liraglutide

Liraglutide is a long acting analog with 97% homology to human GLP-1. It has an additional 16-carbon fatty acid and a small amino acid-spacer that promotes reversible binding to albumin and enhances resistance to DPP- 4 degradation, providing a half-life of approximately 13h [38]. The roles of liraglutide were reported in a lot of clinical trials.

A series of phase 3 clinical trials has examined the efficacy and safety of liraglutide over time periods ranging in duration from 26 to 52 weeks [39]. It was including 5796 patients with type 2 diabetes. Liraglutide, administered as monotherapy or in combination with various oral antidiabetic drugs (OADs), was compared with insulin glargine, exenatide, sitagliptin, glimepiride and various combinations of glimepiride, metformin and rosiglitazone. Liraglutide is effective at improving indices of glycemic control, and has a good tolerability and safety profile. Beneficial effects on HbA1c (mean reduction of 1–1.5%), weight (mean reduction of 1–3.4 kg) and blood pressure (systolic blood pressure decreased by 2.1–6.7 mmHg) are also observed.

A series of phase 3 clinical trialshas examined the efficacy and safety of liraglutide over time periods ranging in duration from 26 to 52 weeks [39]. It was including 5796 patients with type 2 diabetes. Liraglutide, administered as monotherapy or in combination with various oral antidiabetic drugs(OADs), was compared with insulin glargine, exenatide, sitagliptin, glimepiride and various combinations of glimepiride, metformin and rosiglitazone. Liraglutide is effective at improving indices of glycemic control, and has a good tolerability and safety profile. Beneficial effects on HbA1c (mean reduction of 1–1.5%), weight (mean reduction of 1–3.4 kg) and blood pressure (systolic blood pressure decreased by 2.1–6.7 mmHg) are also observed.

In a 20-week prospective multicentre study, 564 nondiabetic obese individuals (31% of whom had prediabetes) were randomized to receive either liraglutide or placebo [40]. Sixty-one percent of the individuals in the liraglutide groups lost at least 5% of body weight from baseline, which was significantly more than the placebo group, and the systolic/diastolic blood pressure was reduced by 5.7/3.7 mmHg. The incidence of metabolic syndrome was reduced by more than 60% in those treated with liraglutide 2.4 mg and 3.0 mg. The prevalence of prediabetes was decreased by 84-96% with liraglutide. Fasting insulin levels initially increased, but as body weight and glucose concentrations gradually decreased, insulin levels were reduced, suggesting the glucose-depended activity of liraglutide on insulin secretion. The two-year results from the extension of the above 20-week trial were reported recently [41]. Estimated weight loss of 7.8 kg and mean systolic blood pressure reduction of 12.5 mmHg was sustained with liraglutide in completers from screening. Between 52%-62% of liraglutide-treated individuals with prediabetes at randomization achieved NGT after two years. The two year prevalence of prediabetes and metabolic syndrome in the liraglutide group was decreased by 52% and 59% respectively.

The multi-national SCALE - Obesity and Pre-diabetes 3, a trial showed that after 56 weeks of treatment, liraglutide 3 mg, in combination with diet and exercise, provided significantly greater weight loss of 8% from baseline, compared to 2.6% with placebo ($p < 0.0001$) [42]. 64% of patients lost more than 5% of their body weight (27% for placebo, $p < 0.0001$). There was also reduction in waist circumference, improvement in lipid profile and blood pressure reduction of 2.82 mmHg (systolic) and 0.89 mmHg (diastolic).

Recently, a 14-week double blind, randomized placebo-controlled study was launched in order to investigate the possible role of liraglutide 1.8 mg treatment in

68 older overweight/obese individuals with prediabetes [43]. More individuals in the liraglutide arm finally lost 7% of baseline weight compared to the placebo group (54% *vs.* 4%). Weight loss after liraglutide therapy was associated with significant reduction of insulin resistance. Steady state plasma glucose concentrations were reduced by 29% in the liraglutide arm compared with no change in the placebo group. In addition, FPG, systolic blood pressure, and triglyceride levels were also significantly decreased in the liraglutide group compared to the placebo group respectively.

FDA has approved Saxenda® (liraglutide 3 mg), the first once-daily human glucagon-like peptide-1 (GLP- 1) analogue for the treatment of obesity.

Individual patient data meta-analysis of LEAD program was performed. Of 4442 patients analysed, 2241 patients had an abnormal ALT at baseline [mean ALT 33.8(14.9) IU/L in females; 47.3(18.3) IU/L in males] [44]. Liraglutide 1.8 mg reduced ALT in these patients *vs.* placebo (8.20 *vs.* 5.01 IU/L; p = 0.003). Adverse effects with 1.8 mg liraglutide were similar between patients with and without baseline abnormal ALT.

LEAN (Liraglutide Efficacy and Action in NASH) trial was a multicentre, double-blinded, randomised, placebo-controlled phase 2 trial to assess subcutaneous injections of liraglutide (1.8 mg daily) compared with placebo for patients who are overweight and show clinical evidence of non-alcoholic steatohepatitis [45]. After 48 weeks treatment, nine (39%) of 23 patients who received liraglutide and underwent end-of-treatment liver biopsy had resolution of definite non-alcoholic steatohepatitis compared with two (9%) of 22 such patients in the placebo group. Two (9%) of 23 patients in the liraglutide group versus eight (36%) of 22 patients in the placebo group had progression of fibrosis. Most adverse events were mild to moderate in severity and transient.

The cardiovascular safety of liraglutide will be prospectively evaluated in the international LEADER trial [46], which is enrolling 9000 patients with type 2 diabetes and a broad range of cardiovascular risk, including known coronary heart disease. Patients will be randomised 1:1 to liraglutide or placebo and will be followed for up to 5 years for adjudicated macrovascular events including non-fatal MI, stroke, and cardiovascular death. The overt result might be released in 2016.

IDegLira

IDegLira is a fixed ratio of insulin degludec (100 U/mL) and liraglutide (3.6 mg/mL) with a maximum dose of 50 Units IDeg/1.8 mg liraglutide, corresponding with the maximum approved dose of liraglutide, where the unit of measure for this fixed-ratio combination will be noted as 'dosing steps'. The combination has the potential to provide improved overall glycemic control whilst mitigating some of the common side effects experienced with GLP-1RAs and basal insulin (*e.g.*, nausea, weightgain, and hypoglycemia).

Two phase 3a clinical trials of IDegLira have been performed, the first being a 26 weeks, treat-to-target, randomized ,open label study comparing IDegLira with insulin degludec or liraglutide alone in insulin-naive patients previously treated with metformin with or without pioglitazone . Treatment with IDegLira produced a significantly greater reduction in HbA1c (-1.9% from baseline) than either degludec (-1.4% from baseline) or liraglutide (-1.3% from baseline) alone after 26 weeks. In addition, a significantly greater proportion of patients achieved glycemic targets of HbA1c< 7% after 26 weeks of treatment with IDegLira than with degludec (81% *vs.* 65%,$p<0.0001$) or liraglutide (60%, $p<0.0001$) and HbA1c <6.5%compared with degludec (70% *vs.* 47%, $p<0.0001$) or liraglutide (70% *vs.* 41%, $p<0.0001$). A mean bodyweight reduction of -0.5 kg with IDegLira,compared with a weight increase of 1.6 kg with degludec and a weight loss of 3.0 kg with liraglutide. In addition, IDegLira also demonstrated a 32% lower rate of

hypoglycemia than degludec despite a lower end-of-trial HbA1c (6.4% *vs.* 6.9%). The incidence of nausea was lower with IDegLira than with liraglutide (9% *vs.* 20% patients).

The second study was a double-blind trial of IDegLira compared with insulin degludec inpatients previously treated with basal insulin [47]. Aspart of the study design in DUAL II, the degludec comparator arm was capped at 50 dose units. A significantly greater reduction in HbA1c compared with those on degludec (capped at 50 Units) after 26 weeks. At the 26-week endpoint, 60% of participants in the IDegLira group had achieved HbA1c<7% *vs.* 23% in the degludec arm and a significantly higher proportion (40%) of patients in the IDegLira arm achieved HbA1c<7% with no confirmed hypoglycemic episodes during the last 12 weeks of treatment and with no weight gain, than in the degludec group. A mean weight loss of 2.7 kg compared with no weight change with degludec. The incidence of nausea was higher in the IDegLira group than in the degludec group in both trials.

Semaglutide

Semaglutide was a once-weekly GLP-1 analog. It is similar to Semaglutide has two amino acid substitutions compared to human GLP-1 (Aib(8), Arg(34)) and is derivatized at lysine 26 [48]. The affinity for the GLP-1 R for semaglutide was three fold less than that of liraglutide, but the albumin binding was increased providing a longer duration of action. In a study of mini-pigs, the plasma half-life was 46.1 h after intravenous administration, and mean residence time was 63.6 h after subcutaneous dosing.

In a Phase II trial, Semaglutide, given subcutaneously once weekly without dose escalation (0.1 – 0.8 mg) or with dose escalation (0.4 mg steps to 0.8 or 1.6 mg E over 1 –2 weeks), was compared with open-label liraglutide once daily (1.2 or 1.8 mg), or placebo over 12 weeks [49]. The group of Semaglutide showed dose-dependent reductions in HbA1c by up to −1.7%, and in bodyweight by up to −4.8 kg. The adverse events of nausea, vomiting and withdrawal due to gastrointestinal symptoms also increased with dose. HbA1c level and weight reductions with semaglutide 1.6 mg were greater than those with liraglutide 1.2 and 1.8 mg.

Lixisenatide

Lixisenatide was developed from exenatide with the C-terminal end modified by addition of six Lys residues, which provides resistance to DPP4 degradation. It was approved in Europe and Japan in 2013 and is given subcutaneously in doses of 10 – 20 µg once daily [50].

In a randomized, open-label, parallel-group, multicentre study, patients (mean HbA1c 7.3%) received subcutaneous lixisenatide QD (10 µg weeks 1–2, then 20 µg; n=77) or liraglutide QD (0.6mg week 1,1.2 mg week 2, then 1.8mg; n=71) 30 min before breakfast [51]. After 28-day treatment, Changes in fasting plasma glucose were greater with liraglutide (-0.3 *vs.* -1.3 mmol/l, $p < 0.0001$). But Lixisenatide reduced PPG significantly more than liraglutide. Change in maximum PPG excursion was -3.9 mmol/l *vs.* -1.4 mmol/l, respectively. Mean HbA1c decreased in both treatment groups (from 7.2% to 6.9% with lixisenatide *vs.* 7.4% to 6.9% withLiraglutide, $p < 0.01$ for the difference betweengroups) as did body weight (-1.6 kg *vs.* -2.4 kg, respectively). Lixisenatide provided greater decreases in postprandial glucagon, insulin and C-peptide. No serious events or hypoglycaemia reported in both groups.

The ELIXA trial was a multicentre, randomized, double-blind study to assess the effect of lixisenatide on cardiovascular outcomes in patients with diabetes and acute coronary disease [52]. Patients with type 2 diabetes who experienced an acute coronary event within 180 days of screening were enrolled into the study. Participants were randomly assigned to receive a subcutaneous injection of lixisenatide (starting dose of 10 µg, increasing to a maximum dose of 20 µg) or volume-matched placebo once daily, in addition to conventional therapy. The trial was powered to determine whether lixisenatide was noninferior or superior to placebo using the primary composite end point of cardiovascular death, myocardial infarction, stroke, or hospitalization for unstable angina. The incidence of hypoglycaemia, hyperglycaemia, pancreatitis, systemic allergic reactions, as well as other safety end points were also assessed. In total, 6,068 patients from 49 countries were enrolled into the study and followed up for a median of 25 months. A primary end point event occurred in 13.4% ($n = 406$) of the lixisenatide group and in 13.2% ($n = 399$) of the placebo group. Statistical analyses of the primary end point events indicated that lixisenatide was noninferior to placebo, but not superior. Treatment with lixisenatide was not associated with an increased occurrence of severe adverse events, such as hypoglycaemia or pancreatic neoplasms.

Albiglutide

Albiglutide is generated in the yeast species Saccharomycescerevisiae by recombinant DNA technology and the resulting therapeutic fusion protein consists of two sequential copies of modified human GLP-1 linked to human albumin.

Albiglutide is a once-weekly, glucagon-like peptide-1 receptor agonist approved during 2014 in both the US and Europe for the treatment of adults with type2 diabetes. The recommended dose is 30 mg with the possibility of up titration to 50 mg based on individual glycemic response.

The eight Phase III clinical trials have provided evidence that albiglutide in monotherapy and as an add-on to different background therapies confers placebo-corrected reductions in glycemia with changes inglycated hemoglobin of -0.8 to -1.0% [53]. Albiglutide did not cause significant weight loss compared to placebo, but the adverse events profile was favorable with gastrointestinal adverse events occurring only slightly more with albiglutide than placebo.

A prospective meta-analysis of the cardiovascular safety of albiglutide was performed [54]. It was a meta-analysis of eight phase 3 trials and one phase 2b trial in which patients were randomly assigned to albiglutide, placebo, or active comparators (glimepiride, insulin glargine, insulin lispro, liraglutide, pioglitazone, or sitagliptin). The safety population included 5107 patients, of whom 2524 took albiglutide (4870person-years) and 2583 took comparators (5213 person-years). The primary endpoint was a composite of first occurrence of major adverse cardiovascular events (*i.e.*, cardiovascular death, non-fatal myocardial infarction, or non-fatal stroke) or hospital admission for unstable angina. Secondary endpoints were major adverse cardiovascular events alone, all-cause mortality, silent myocardial infarction, hospital admission for heart failure, chest pain, other angina, and subdural or extradural haemorrhage. The primary endpoint was not significantly different between albiglutide and all comparators. Major adverse cardiovascular event alone was also not significantly different. When albiglutide was compared separately with placebo or active comparators, we noted no significant differences. We detected no significant differences in the other secondary endpoints. More patients had atrial fibrillation or atrial flutter in the albiglutide group than in the all-comparators group. So we got the conclusion that cardiovascular events were not significantly more likely to occur with albiglutide than with all comparators.

Dulaglutide

Dulaglutide is a once-weekly, GLP-1receptor agonist, consisting of a DPP-IV-protected GLP-1analogue covalently linked to a human IgG4-Fc heavy chain by a small peptide linker. Dulaglutide as a once-weekly subcutaneous injection drug, adjunct to diet and exercise, for the treatment of adult patients with type 2 diabetes mellitus was approved in US and EU in 2014.

The efficacy and tolerability of once-weekly subcutaneous dulaglutide 0.75 or 1.5 mg in patients with type 2 diabetes has been evaluated in the AWARD (Assessment of Weekly Administration of dulaglutide) programme of clinical studies [55]. The studies were all randomized trials versus active comparators in patients with type 2 diabetes who were treatment-naive or had inadequate control with current diabetic regimens.

Indeed, the effects of GLP-1R agonists on insulin secretion are not a simple phenomenon. These medications can increase insulin secretion in a glucose-depended manner after acting directly on the β cell. They can also decrease insulin secretion secondary to weight loss and enhancement of insulin sensitivity.

THE ROLES OF DPP-4 INHIBITORS

Native GLP-1 is rapidly inactivated (half-life of 1-2 min) by the ubiquitously expressed proteolytic enzyme DPP-4 [56]. The DPP-4 inhibitors are a class of oral antidiabetic agents that improve glycemic control by increasing both GLP-1 and GIP concentrations [57]. Studies demonstrated that DPP-4 inhibitors have trophic effects on pancreatic β-cells. Specifically they enhance β-cell proliferation, regeneration and differentiation; thus they increase β-cell mass. They also inhibit β-cell apoptosis, including human β-cells, through inhibition of the caspase pathway [58].

A recent meta-analysis [59] showed that DPP-4 inhibitors were superior to placebo in reducing HbA1c levels in adults with T2DM taking at least two oral agents. DPP-4 inhibitors offer various advantages when compared to other glucose-lowering agents. Despite they have been commercialized since a few years only, available data obtained in randomised controlled trials are of better quality compared to those available with ancient classical glucose lowering agents, especially in more fragile populations such as elderly people, individuals with renal impairment or at high cardiovascular risk and at higher risk of hypoglycaemia. However, there remain uncertainties and controversies that should be resolved by further ongoing large prospective controlled trials and increasing clinical experience combined with a careful post-marketing surveillance.

The DPP-4 inhibitors are approved for use in patients with T2DM in many regions of the world, including sitagliptin, vildagliptin, saxagliptin, linagliptin and alogliptin. Thus, the DPP-4 inhibitors occupy an increasing place in the anti-diabetes drugs and offer new opportunities for a personalized medicine in patients with T2DM. The DPP-4 inhibitors as monotherapy have an almost comparable glucose-lowering effect to that of a thiazolidinedione or a sulphonylurea, with the advantage of no weight gain [60]. The improvement in albuminuria seen with DPP-4 inhibitors suggests that these anti-diabetes drugs may potentially provide renal benefits beyond their glucose-lowering effects [61]. However, metformin therapy is much cheaper than treatment with any DPP-4 inhibitor so that metformin should be considered as first-line drug in patients with T2DM. A DPP-4 inhibitor may be an alternative only in case of metformin intolerance or contraindication [34].

Oral DPP-4 inhibitors are well-suited medications in older patients because of their consistent efficacy, low risk of hypoglycaemia and ease of use. Initial studies suggested that for elderly patients with T2DM, reductions in HbA1c after treatment with a DPP-4 inhibitor were not significantly different from those in younger patients and that the use of DPP-4 inhibitors was associated with a low risk of hypoglycaemia and also weight neutrality. These medications have proven to be quite popular given their ease of dosing (one pill a day), lack of hypoglycemia, and weight gain improvement in glycemic control associated with their use. To date, no serious adverse events have emerged even though there are several large ongoing studies assessing long-term cardiovascular safety.

Sitagliptin

Sitagliptin is the first DPP-4 inhibitor introduced in clinical practice [62]. In IFG individuals, sitagliptin did not alter fasting but increased postprandial intact GLP-1 concentrations, while total postprandial GLP-1 concentrations were reduced.

However, both fasting and postprandial glucose values were unchanged with sitagliptin therapy [63]. A four week open-label, parallel group study investigated the effects of sitagliptin on insulin secretion and endogenous glucose production in individuals with IFG [64]. While in another trial, treatment with sitagliptin resulted in a small but significant decrease in FPG compared to baseline. However, endogenous glucose production, insulin sensitivity and β cell response indices were unchanged after 4 weeks of sitagliptin therapy. Moreover, a 12-week multicentre, double-blind, randomized, parallel group study investigated the effects of sitagliptin 100 mg daily compared to placebo in 71 patients with acute coronary syndrome having IGT or T2DM [65]. Insulinogenic index increased significantly in the sitagliptin group compared to the placebo group. The acute insulin response to glucose increased significantly in the sitagliptin group compared to the placebo. It was speculated that the limited ability of DPP-4 inhibitors to increase insulin secretion in IFG could be due to their glucose depended mechanism, since glucose concentrations are only modestly elevated in IFG. It was suggested that if sitagliptin actions could extend to human prediabetics, then sitagliptin might delay the onset of diabetes.

The 14 studies of sitagliptin as monotherapy were reviewed in 2008 [66] and summarized in 2009 [67]. In general, sitagliptin reduced HbA1c by 0.5–0.7 % from baseline level of ~8 %. Sitagliptin for the treatment of NAFLD with type 2 DM was safe and showed similar antidiabetic effects as reported for type 2 DM, suggesting that tight glycemic control would contribute to the improvement of NAFLD based from the findings of correlation between the changes of HbA1c and transaminases [68]. Therefore, sitagliptin is an effective antidiabetic drug in T2DM, especially in obese individuals.

Vildagliptin

Vildagliptin was approved for clinical use for patients with T2DM in 2007 and is available in over 100 countries. Vildagliptin is an oral agent that inhibits DPP-4 and increases both active GLP-1 and GIP levels. The long-term administration of vildagliptin improved first-phase insulin secretion and insulin sensitivity index suggesting that this drug has the capacity to repair impairments in pancreatic β-cell function and insulin resistance in type 2 diabetes [69]. A single administration of vildagliptin attenuated postprandial endothelial dysfunction and postprandial hypertriglyceridemia, suggesting that vildagliptin may be a promising antiatherogenic agent [70]. Suppression of glucagon release by vildagliptin may improve glycemic control without increasing insulin levels in patients with type 2 diabetes [71].

In 22 IFG individuals, treatment with vildagliptin resulted in a slight increase in fasting GIP but not GLP-1 levels, while marked increases of both intact GLP-1 and GIP levels during a meal tolerance test were reported [72]. The disposition index (DI) was increased by 69% and insulin sensitivity by 25%, suggesting an improvement of β-cell function would be expected to occur. In a study with IGT individuals, there was a marked and sustained increase in active GLP-1 and GIP levels in vildagliptin group compared to the placebo [73], and these effects were associated with significant improvements in β -cell function and α-cell function. In addition, a three month, double-blind, placebo-controlled study was organized in a population of 48 stable renal transplant recipients, at least six months after transplantation, with newly diagnosed IGT [74]. Participants were randomized to receive vildagliptin, pioglitazone or placebo. The HbA$_{1c}$ and the 2h plasma glucose reduction were statistically significant between treatment groups and placebo. Therefore, vildagliptin is useful in prediabetes patients.

It is indicated as monotherapy and combination treatment with metformin, sulfonylurea, thiazolidinedione, and insulin, and it improved glycemic control in T2DM patients [75]. A large observational study of 45,868 patients with T2DM across 27 countries assessed the effectiveness and safety of vildagliptin. In this real life, treatment with vildagliptin was associated with a higher proportion of patients with T2DM achieving better glycemic control without tolerability issues in the Middle East [76]. Moreover, in real-life clinical practice in Bulgaria, vildagliptin is associated with a greater HbA$_{1c}$ drop, and a higher proportion of patients reaching target HbA$_{1c}$ without hypoglycemia and weight gain compared to comparator [77]. The monotherapy studies showed significant HbA$_{1c}$ lowering by vildagliptin [78]. Vildagliptin monotherapy was also tested *vs.* alpha glucosidase inhibitors in China. After 24 weeks, vildagliptin lowered A$_{1c}$ by 1.4 % (baseline of 8.6 %), similar to the result seen with acarbose [79]. Among Japanese patients, vildagliptin was more powerful than voglibose after 12 weeks.

Vildagliptin lowered A_{1c} by 0.95 % from baseline of 7.6% [80]. Patients treated with metformin and vildagliptin showed better adherence and metabolic control and lower rates of hypoglycemia, resulting in lower health care costs for the national health system [81]. The combination therapy of vildagliptin and nateglinide is effective and safe in Japanese type 2 diabetes, and the improved glycemic control is as a result of augmentation of nateglinide-induced early phase insulin secretion [82]. The combination therapy of vildagliptin plus an α-glucosidase inhibitor effectively reduced the fasting blood glucose, postprandial glucose and HbA1c levels in patients without inducing weight gain or hepatorenal dysfunction [83].

Therefore, treatment with metformin plus vildagliptin compared with metformin plus sulphonylurea is expected to result in a lower incidence of diabetes-related adverse events and to be a cost-effective treatment strategy [84].

Saxagliptin

Saxagliptin improved glycaemia and prevented the reduction in fasting homeostatic model 2 assessment of β-cell function values. Saxagliptin may reduce the usual decline in β-cell function in T2DM, thereby slowing diabetes progression [85]. Saxagliptin is recommended as add-on therapy to metformin and as part of two- or three-drug combinations in patients not meeting individualized glycemic goals with metformin alone or as part of a dual-therapy regimen [86]. A recent review showed that saxagliptin was generally well tolerated and consistently improved glycemic control, as assessed by reductions from baseline in HbA1c, fasting plasma glucose concentration, and postprandial glucose concentration, regardless of the presence or absence of baseline cardiovascular disease, hypertension, statin use, and number of cardiovascular risk factors [87]. Saxagliptin improves glycemic control and is generally well tolerated in patients with T2DM, irrespective of concomitant statin therapy [88].

In a dose-finding study, saxagliptin was used at 2.5–40 mg a day for 12 weeks in drug-naive patients with HbA1c 6.8–9.7 % [89]. Treatment resulted in HbA1c reduction ~0.8 % with maximum efficacy reached at the 5 mg dose. Saxagliptin as monotherapy was also assessed at three strengths (2.5, 5, and 10 mg daily) in a 24-

week trial of 401 uncontrolled patients and resulted in decreased A1c (from baseline mean 7.9 %) by 0.43, 0.46, and 0.54 %, respectively [90]. As is typically observed, the higher the baseline HbA1c, the larger absolute change in HbA1c was seen in saxagliptin studies. Thus, when the patients with baseline HbA1c of 9.5 % were studied for 24 weeks, 10 mg dose led to 1.7 % HbA1c improvement [91].

A recent study showed that compared with acarbose-metformin, saxagliptin-metformin was more effective in glucose control with similar glycaemic variability [92]. In patients with T2DM in China, saxagliptin+metformin was more cost-effective and was well tolerated with fewer adverse effects compared with glimepiride+metformin [93]. As a second-line therapy for T2DM, saxagliptin may address some of the unmet medical needs attributable to adverse effects in the treatment of T2DM. The incidence of confirmed/severe hypoglycaemia was lower with saxagliptin compared to glimepiride. As avoiding hypoglycaemia is a key clinical objective in elderly patients, saxagliptin is a suitable alternative to glimepiride in patients with T2D aged ≥ 65 years [94]. Saxagliptin add-on therapy was generally well tolerated in older patients aged ≥ 65 years with type 2 diabetes mellitus, with a long-term safety profile similar to that of placebo [95]. Recently, investigators in the SAVOR-TIMI (Saxagliptin Assessment of Vascular Outcomes Recorded in Patients with Diabetes Mellitus-Thrombolysis in Myocardial Infarction) 53 study suggested that DPP-4 inhibition with saxagliptin did not increase or decrease the rate of ischemic events, although the rate of hospitalization for heart failure was increased [96]. Although saxagliptin improves glycemic control, other approaches are necessary to reduce cardiovascular risk in patients with diabetes.

Linagliptin

Linagliptin was typically studied at 5 and 10 mg daily dose. In the US, the 5 mg became the only marketed dose. Linagliptin 5 mg once-daily and 2.5 mg twice-daily provided bioequivalent exposure and similar inhibition of DPP-4 over the whole dosing interval [97]. In T2DM patients, multiple rising doses of linagliptin were well tolerated and resulted in significant improvements of glucose parameters [98]. Monotherapy with linagliptin produced a significant, clinically meaningful and sustained improvement in glycaemic control, accompanied by enhanced parameters of β-cell function [99]. In two examples of monotherapy studies, 5 mg linagliptin reduced A_{1c} by 0.44 % after 24 weeks [99] and by 0.49 %

after 12 weeks and by 0.44 % at 26 weeks among 561 Japanese patients [100]. Linagliptin is more effective than glimepiride at achieving a composite outcome of target HbA_1c < 7% with no hypoglycaemia and no weight gain over 2 years [101]. Therefore, linagliptin is a safe and effective glucose-lowering treatment in T2DM patients with moderate-to-severe renal impairment for whom sulphonylurea treatment is no longer sufficient [102, 103].

Linagliptin was an effective, well-tolerated treatment in subjects with T2DM and insufficient glycemic control, both as monotherapy or added-on to metformin/metformin plus sulfonylurea [104]. Initial combination of linagliptin and metformin was well tolerated over the 1-year extension period, with low risk of hypoglycaemia, and improved glycaemic control compared with metformin alone [105]. Linagliptin added to basal insulin therapy also significantly improved glycemic control relative to placebo without increasing hypoglycemia or body weight [106]. In elderly patients with type 2 diabetes linagliptin was efficacious in lowering glucose with a safety profile similar to placebo [107].

Linagliptin in combination with metformin, sulphonylurea and has a favourable safety profile and is an efficacious and well tolerated treatment option for Chinese patients with inadequately controlled T2DM. Reduction of sulphonylurea dose should be considered to minimise risk of hypoglycaemia [108]. Once-daily linagliptin showed safety and tolerability over 1 year and provided effective add-on therapy leading to significant HbA1c reductions, similar to metformin, over 52 weeks in Japanese patients [109]. Initial combination therapy with linagliptin plus pioglitazone was well tolerated and produced significant and clinically meaningful improvements in glycaemic control [110]. This combination may offer a valuable additive initial treatment option for T2DM, particularly where metformin either is not well tolerated or is contraindicated, such as in patients with renal impairment. Linagliptin demonstrated a favorable safety profile in healthy Chinese volunteers, with a pharmacokinetic profile that was similar to that observed previously in subjects of Japanese, Caucasian, or African American origin [111].

Alogliptin

Alogliptin is the newest DPP-4 inhibitor approved for T2DM therapy, either alone or in combination with other antidiabetic agents [112]. Alogliptin monotherapy was well tolerated and significantly improved glycemic control in patients withT2DM, without raising the incidence of hypoglycemia [113]. Alogliptin was studied usually at 12.5 or 25 mg daily doses. In a meta-analysis of 10 double-blind studies, the 12.5 mg dose led to 0.81 % and 25 mg to 0.98 % reduction in A_{1c} [114]. In Japanese patients, alogliptin after 12 weeks reduced A_{1c} by 0.68 % at 12.5 mg dose and by 0.77 % at 25 mg daily [115]. Finally, in newly diagnosed drug-naïve subjects with very high initial A_{1c} (10.5 %), 25-mg alogliptin led to 1.77 % reduction after 12 weeks [116].

A 26-week, double-blind, parallel-group study showed that alogliptin plus pioglitazone combination treatment appears to be an efficacious initial therapeutic option for type 2 diabetes [117]. Moreover, addition of alogliptin to pioglitazone therapy also significantly improved glycemic control in patients with type 2 diabetes and was generally well tolerated [118]. Another study showed that alogliptin is an effective and safe treatment for T2DM when added to metformin for patients not sufficiently controlled on metformin monotherapy [119]. Adding alogliptin to an existing metformin–pioglitazone regimen provided superior glycaemic control and potentially improved β-cell function versus uptitrating pioglitazone in T2DM patients, with no clinically important differences in safety [120]. In patients with T2DM inadequately controlled by glyburide monotherapy, the addition of alogliptin resulted in clinically significant reductions in HbA1c without increased incidence of hypoglycaemia [121]. Moreover, adding alogliptin to previous insulin therapy (with or without metformin) significantly improved glycaemic control in T2DM inadequately controlled on insulin, without causing weight gain or increasing the incidence of hypoglycaemia [122].

A multicentre, double-blind, active-controlled study randomized 2639 patients aged 18–80 years to 104 weeks of treatment with metformin in addition to alogliptin 12.5 mg once daily, alogliptin 25 mg once daily or glipizide 5 mg once daily, titrated to a maximum of 20 mg [123]. The results showed that HbA1c reductions at week 104 were −0.68%, −0.72% and −0.59% for alogliptin 12.5 and 25 mg and glipizide, respectively. Fasting plasma glucose concentration decreased by 0.05 and 0.18 mmol/l for alogliptin 12.5 and 25 mg, respectively. Mean weight changes were −0.68, −0.89 and 0.95 kg for alogliptin 12.5 and 25 mg and glipizide, respectively. Therefore, alogliptin efficacy was sustained over 2 years in patients with inadequate glycaemic control on metformin alone. Moreover, in the EXAMINE study [124], the HbA1c levels were significantly lower with alogliptin than with placebo. Incidences of hypoglycemia, cancer, pancreatitis, and initiation of dialysis were similar with alogliptin and placebo. Therefore, among patients with type 2 diabetes who had a recent acute coronary syndrome, the rates of major adverse cardiovascular events were not increased with the DPP-4 inhibitor alogliptin as compared with placebo. Therefore, alogliptin significantly improved glycaemic control in T2DM, without causing weight gain or increasing the incidence of adverse cardiovascular events.

WEIGHT-LOWERING PROPERTIES OF GLP-1 AGONISTS AND DPP-4 INHIBITORS

Among patients with T2DM, abdominal obesity is a very common complication [125]. As patients with T2DM have an inherent elevated risk of cardiovascular complications [126], the combination of abdominal obesity would greatly increase the risk of cardiovascular morbidity such as myocardial infarction and stroke as well as total mortality [127]. Waist circumference is widely used as a tool for estimation of abdominal obesity and visceral fat accumulation clinically. Improvement of abdominal obesity and visceral fat accumulation was considered to be associated with reduction of major cardiovascular risk factors [128]. Therefore, it is important to better control body weight, particularly waist circumference, for patients with T2DM to achieve favorable cardiovascular safety profiles.

Whereas DPP-4 inhibitors are almost weight neutral (or may be associated with a marginal weight loss), GLP-1 receptor agonists are consistently associated with a moderate weight reduction (on average, 2-4 kg) [129]. Some GLP-1 receptor agonists, especially liraglutide—1.8 mg once daily and liraglutide—1.2 mg once daily, were associated with a significant reduction in waist circumference. Moreover, a systematic review [130] showed that significant reductions on waist circumference following treatment of liraglutide-1.8 mg once daily (-5.24 cm), liraglutide-1.2 mg once daily (-4.73 cm) and exenatide-10 ug twice daily (-1.34 cm,) were detected versus placebo. Therefore, the GLP-1 receptor agonists and DPP-4 inhibitors are the better choice for the T2DM with obesity.

SAFETY

An acceptable safety profile is of major importance for every intervention administered in order to prevent or delay T2D. The most common adverse effects GLP-1R agonists are gastrointestinal, including nausea, vomiting and diarrhea

[131]. However, they occur early on during treatment and tend to be transient. For the DPP-4 inhibitors, adverse effects resemble that of placebo, with nasopharyngitis and headache being the most common described [132]. Moreover, discontinuation of therapy because of side effects was similar to placebo [133].

Hypoglycemia is a major concern for clinicians and their patients in management of type 2 diabetes. Hypoglycemia is an infrequent complication of the incretin-related medications in T2DM. In the ACCORD study, a benchmark long-term outpatient trial, intensive treatment was associated with a 2.5-fold increase in hypoglycemic events [134]. Because of the glucose-dependent mechanism, the incretin-based therapies when used alone or with metformin have a low risk of hypoglycemia. Furthermore, DPP-4 inhibitors do not suppress glucagon release in response to hypoglycemia [135].

Current studies suggested a probable role of GLP-1 receptor agonists on the development of pancreatic cancer and thyroid cancer in rodents, but such an effect in humans is not remarkable due to the lower or lack of expression of GLP-1 receptor on human pancreatic ductal cells and thyroid tissues [136]. However, findings in human studies are controversial and inconclusive. A recent study showed that some parts of papillary thyroid carcinomas (PTC) tissues express GLP-1 receptor, and the GLP-1 receptor expression in PTC was negatively correlated with tumor multifocality [137]. Both liraglutide and exenatide were shown to promote the development of thyroid C cell cancer after chronic therapy in rodents [138]. And an elevated risk for thyroid carcinoma was found in one study [139]. However, thyroid C cells in humans and monkeys express lower levels of GLP-1receptors [140]. Long-term treatment with high doses liraglutide did not produced thyroid C cell proliferation in monkeys, while no association between calcitonin levels and liraglutide, up to 3 mg daily, was established in large numbers of patients with T2DM [141]. Some preliminary studies even suggested a potentially beneficial effect on the development of other cancers with the use of incretins. Therefore, continuous monitoring of the cancer issues related to incretin based therapies is required, even though the benefits may outweigh the potential cancer risk in the general patients with type 2 diabetes mellitus.

Two recent meta-analysis suggested that the DPP-4 inhibitors may have a neutral effect or reduce the risk of cardiovascular events and all-cause mortality in patients with T2DM [142, 143]. Moreover, other two recent meta-analysis reported that GLP-1receptor agonists do not appear to increase cardiovascular morbidity in comparison with placebo or other active drugs [144, 145]. In addition, other two meta-analysis suggested a possible increased risk of developing heart failure after DPP-4 inhibitors therapy [145, 146]. A recent meta-analysis of all randomised controlled trials with DPP-4 inhibitors showed no cardiovascular benefit (or harm) with DPP-4 inhibitors [146]. However, results of posthoc analyses of Phase II/III controlled trials suggest a possible cardioprotective effect with a trend toward lower incidence of major cardiovascular events with sitagliptin, vildagliptin, saxagliptin, linagliptin, or alogliptin, compared with placebo or other active glucose-lowering agents [147, 148]. The SAVOR-TIMI 53 trial showed that DPP-4 inhibition with saxagliptin does not increase or decrease the rate of ischemic events, though the rate of hospitalization for heart failure was increased [149]. The EXAMINE trial showed that among patients with T2DM who had a recent acute coronary syndrome, the rates of major adverse cardiovascular events were not increased with alogliptin as compared with placebo, confirming the cardiovascular safety of the DPP-4 [124]. The recent TECOS study found that among patients with type 2 diabetes and established cardiovascular disease, sitagliptin did not appear to increase the risk of major adverse cardiovascular events, hospitalization for heart failure, or other adverse events [150]. Currently, a large number of long-term cardiovascular outcome trials in patients with T2D are being performed in order to clarify the cardiovascular safety and efficacy of incretin-based therapies [151].

Small preclinical studies and some post-marketing reports raised the possibility of an increased risk of pancreatitis with incretin-based therapies [152]. A trend towards a slightly elevated risk of pancreatitis, only with GLP-1 receptor agonists, was also shown in a recent pooled analysis of phase III trials, although the number of cases was very small and the statistical power was limited [153]. In a case-control study based upon an administrative database of US adults with T2DM, treatment with the GLP-1-based therapies (sitagliptin and exenatide) was indeed associated with increased odds of hospitalization for acute pancreatitis [154]. Pancreatitis is well known to predispose to pancreatic cancer, and there was a signal for cancer of the pancreas for sitagliptin and exenatide in the FDA database [139]. To further support this concern, alarming data have been reported from human pancreatic analyses [155]. Incretin therapy in humans resulted in a marked expansion of the exocrine and endocrine pancreatic compartments, the former being accompanied by increased proliferation and dysplasia and the latter by a-cell hyperplasia with the potential for evolution into neuroendocrine tumours [155]. However, the various data presented have serious methodological deficiencies that preclude any meaningful conclusions [156]. In a recent systematic review and meta-analysis of randomised and non-randomised studies, the available evidence suggests that the incidence of pancreatitis among patients using DPP-4 inhibitors is low and that the drugs do not increase the risk of pancreatitis [157]. It is probably more difficult to have a definite answer regarding the pancreatic cancer concern. According to Nauck [158], there is no case report of pancreatic cancer diagnosed after exposing to DPP-4 inhibitors in a patient in whom there had previously been a morphologically tumour-free pancreas. Because pancreatic carcinomas develop slowly, one would probably not expect to see such a case after at most a few years of treatment. A recent meta-analysis conducted by Monami *et al.* did not find an increased risk of pancreatitis associated with DPP-4 inhibitors [159]. The two recent large multi-center randomized placebo-controlled trials EXAMINE [160] and SAVORTIMI [149] showed no significant difference between the active and placebo groups in acute or chronic pancreatitis. However, larger preclinical studies did not establish an association of incretin-based therapies with pancreatitis [161]. Interestingly, in three studies, GLP-1recepor activation or DPP-4 inhibition had a beneficial effect on exocrine pancreatic function and structure [162, 163]. Moreover, large retrospective population studies and a recent meta-analysis suggested a negative association of incretin-based therapies with either pancreatitis or pancreatic cancer [164-173]. Recently the FDA reevaluated more than 250 toxicology studies, organized in nearly 18000 healthy animals, and found no association with pancreatitis or any pancreatic toxicity. A recent well-controlled population-based study [174] suggests there is no higher short-term pancreatic cancer risk with DPP-4 inhibitor treatment relative to SU and TZD treatment. Moreover, a recent meta-analysis which included nine studies and 1324515 patients did not suggest that incretin-based therapy is associated with acute pancreatitis [175]. Thus, large post marketing surveillance programmes and further carefully analyzed studies are needed to completely resolve the pancreatic safety issues with DPP-4 inhibitors.

No significant changes in liver enzymes were reported with DPP-4 inhibitors alone or in combination with various other glucose-lowering agents, in clinical trials up to 2 years [176]. With the possible exception of vildagliptin, none of the DPP-4 inhibitors has shown hepatotoxicity in clinical trials. Vildagliptin has exhibited a low incidence of increased liver enzymes. Yet hepatic impairment has not affected vildagliptin pharmacokinetics [177]. All DPP-4 inhibitors except for vildagliptin are approved in patients with hepatic insufficiency without dose adjustment. Vildagliptin has been associated with mild elevations in liver transaminases without demonstration of increased rate of actual hepatic adverse events. However, it is recommended that liver function tests are performed before initiation and periodically after starting vildagliptin therapy, and its use is contraindicated in patients with significant liver disease. All the DPP-4 inhibitors except linagliptin are eliminated renally and are reported to accumulate in patients with renal insufficiency. Linagliptin, however, has primarily a non-renal route of excretion and can be used without dose adjustment in patients at all stages of renal disease. All DPP-4 inhibitors are approved for those patients with mild renal insufficiency (typically defined as creatinine clearance ≥ 50 mL/min). Saxagliptin, alogliptin, and sitagliptin can be used (in USA) at reduced dosage in those with moderate and even severe renal disease. Linagliptin is approved for use at the same 5 mg daily dose regardless of renal function.

In a word, the incretin-based therapy is attractive because it offers improved glycemic control without the untoward adverse events such as hypoglycemia, weight gain and severe adverse effects.

CONCLUSION

During the last two decades there has been an immense investigation in order to understand the pathophysiology of the early stage of hyperglycemia, which often progress to overt T2DM within a few years, as β-cell decline and failure progresses. The huge burden resulting from the complications of T2DM raised the need of novel therapeutic strategies to prevent its development. In the past several years, addressing the "incretin defect" among patients with T2DM has become a popular strategy. The beneficial effects of incretin-based therapies on β-cell function in T2DM patients, together with their strictly glucose-depended mechanism of action, suggested their possible use in individuals with prediabetes, when greater β-cell mass and function are preserved and the possibility of β-cell salvage is higher. Thus, the incretin enhancers were placed as the preferred addition to those patients in whom metformin monotherapy was not sufficient to achieve the desired glycemic targets. In general, these medications, when pitted against sulfonylureas, had similar effect on glycemic control while avoiding the risks of hypoglycemia and weight gain.

Table 1: The efficacy and safety of GLP-1 agonists and DPP-4 inhibitors.

	Efficacy	Safety
GLP-1 agonists	1.HbA1c: 1.0-1.5% above target 2.Fasting plasma glucose:1.1-3.9mmol/L above target 3.Postprandial glucose:>3.3-5.6mmol/L 4.Weight loss 5.Improvement in multiple cardiovascular risk factors 6.Inhibition of inflammatory responses of atherosclerotic plaque macrophages 7. Reduction of systolic/diastolic blood pressure 8. Reduction of prediabetes and metabolic syndrome	1. Gastrointestinal: nausea, vomiting and diarrhea 2. Low risk of hypoglycemia 3. More costly than DPP-4 inhibitors
DPP-4 inhibitors	1.HbA1c: 0.5-1.0% above target 2.Fasting plasma glucose:0-1.7mmol/L above target 3.Postprandial glucose:>3.3mmol/L 4.Improvement of β-cell function 6.Might delay the onset of diabetes 7.Renal benefits beyond their glucose-lowering effects 8.Attenuated postprandial endothelial dysfunction and postprandial hypertriglyceridemia	1. Nasopharyngitis and headache 2. Low risk of hypoglycaemia

The safety of incretin-based therapies remains a topic of scientific discussion and exploration. However, the risk of possible serious adverse effects associated with incretin-based therapies cannot be estimated. Future data from the ongoing clinical studies, which will improve the statistical power of prospective studies and facilitate larger meta-analyses, are crucially anticipated in order to clarify their long term safety. The large, long term, well designed future diabetes prevention trials of incretin-based therapies will be required in order to determine whether they can stabilize or reverse β-cell loss and prevent the development of T2DM.

The efficacy and safety of GLP-1 agonists and DPP4 inhibitors was shown in Table **1**. In addition to safety and efficacy of incretin-based therapies, the cost is another significant issue that must be taken into consideration. Although the cost of incretin-based therapies is greater compared to other glucose-lowering therapies, long term effectiveness of these agents can be associated with a decreased in the cost of management of T2DM and its complications compared to other therapies.

The GLP-1 agonists and DPP4 inhibitors add vital new tools to the doctor's armoury in the fight against T2DM. These options may simply improve treatment flexibility and convenience. We can use incretin therapies to manage patient care effectively and provide tailored treatment regimens that may promote increased adherence to therapy, improve outcomes and potentially slow down T2DM progression.

ACKNOWLEDGEMENTS

Declared none.

CONFLICT OF INTEREST

The author confirms that author has no conflict of interest to declare for this publication.

REFFERENCES

[1]   Editorial. The global challenge of diabetes. Lancet 2008; 371: 1723.
[2]   Tao Z, Shi A, Zhao J. Epidemiological perspectives of diabetes. Cell Biochem Biophys 2015; Feb 25. [Epub ahead of print].
[3]   Wild SH, Farouhi NG. What is the scale of the future diabetes epidemic and how certain are we about it? Diabetologica 2007; 50: 903-5.
[4]   Holst JJ. Glucagon-like peptide-1: from extract to agent. The Claude Bernard Lecture 2005, Diabetologia 2006; 49: 253-60.
[5]   Drucker DJ. The biology of incretin hormones. Cell Metab 2006; 3: 153-65.
[6]   Holst JJ, Vilsboll T, Deacon CF. The incretin system and its role in type 2 diabetes mellitus. Mol Cell Endocrinol 2009; 297(1-2): 127-36.
[7]   Nauck MA, Heimesaat MM, Orskov C, Holst JJ, Ebert R, Creutzfeldt W. Preserved incretin activity of glucagon-like peptide 1 [7-36 amide]but not of synthetic human gastric inhibitory polypeptide in patients with type-2 diabetes mellitus. J Clin Invest 1993; 91(1): 301-7.

[8]   Naslund E, Barkeling B, King N, *et al*. Energy intake and appetite are suppressed by glucagon-like peptide-1 (GLP-1) in obese men. Int J Obes Relat Metab Disord 1999; 23(3): 304-11.

[9]   Flint A, Raben A, Astrup A, Holst JJ. Glucagon-like peptide 1 promotes satiety and suppresses energy intake in humans. J Clin Invest 1998; 101(3): 515-20.

[10]   Qin X, Shen H, Liu M, *et al*. GLP-1reduces intestinal lymph flow, triglyceride absorption, and apolipoprotein production in rats. Am J Physiol Gastrointest Liver Physiol 2005; 288(5) G943-9.

[11]   Nauck MA, Homberger E, Siegel EG, *et al*. Incretin effects of increasing glucose loads in man calculated from venous insulin and C-peptide responses. J Clin Endocrinol Metab 1986; 63: 492-8.

[12]   Knop FK, Vilsboll T, Hojberg PV, *et al*. Reduced incretin effect in type 2 diabetes: cause or consequence of the diabetic state? Diabetes 2007; 56: 1951-9.

[13]   Byrne MM, Gliem K, Wank U, *et al*. Glucagon-like peptide 1 improves the ability of the beta-cell to sense and respond to glucose in subjects with impaired glucose tolerance. Diabetes 1998; 47: 1259-65.

[14]   Singh AK. Incretin response in Asian type 2 diabetes: Are indians different? Indian J Endocrinol Metab 2015; 19(1): 30-8.

[15]   Dhanvantari S, Izzo A, Jansen E, Brubaker PL. Coregulation of glucagon-like peptide-1 synthesis with proglucagon and prohormone convertase 1 gene expression in enteroendocrine GLUTag cells. Endocrinology 2001; 142: 37-42.

[16]   Mayo KE, Miller LJ, Bataille D, *et al*. The glucagon receptor family. Pharmacol Rev 2003; 55: 167-94.

[17]   Buteau J, Foisy S, Joly E, Prentki M. Glucagon-like peptide 1 induces pancreatic beta-cell proliferation via transactivation of the epidermal growth factor receptor. Diabetes 2003; 52: 124-32

[18]   Buteau J, Foisy S, Rhodes CJ, Carpenter L, Biden TJ, Prentki M. Protein kinase Czeta activation mediates glucagon-like peptide-1-induced pancreatic beta-cell proliferation. Diabetes 2001; 50: 2237-43

[19]   Scheen AJ, Esser N, Paquot N. Antidiabetic agents: Potential anti-inflammatory activity beyond glucose control. Diabetes Metab 2015; Mar 17: [Epub ahead of print].

[20]   Göke R, Fehmann HC, Linn T, *et al*. Exendin-4 is a high potency agonist and truncated exendin-(9-39)-amide an antagonist at the glucagon-like peptide 1-(7-36)-amide receptor of insulin-secreting beta-cells. J Biol Chem 1993; 268: 19650-5.

[21]   Blase E, Taylor K, Gao HY, Wintle M, Fineman M. Pharmacokinetics of an oral drug (acetaminophen) administered at various times in relation to subcutaneous injection of exenatide (exendin-4) in healthy subjects. J Clin Pharmacol 2005; 45(5): 570-7.

[22]   Edwards CM, Stanley SA, Davis R, *et al*. Exendin-4 reduces fasting and postprandial glucose and decreases energy intake in healthy volunteers. Am J Physiol Endocrinol Metab 2001; 281(1): E155-61.

[23]   Egan JM, Clocquet AR, Elahi D. The insulinotropic effect of acute exendin-4 administered to humans: comparison of nondiabetic state to type 2 diabetes. J Clin Endocrinol Metab 2002; 87(3): 1282-90.

[24]   Kolterman OG, Buse JB, Fineman MS, *et al*. Synthetic exendin-4 (exenatide) significantly reduces postprandial and fasting plasma glucose in subjects with type 2 diabetes. J Clin Endocrinol Metab 2003; 88(7): 3082-9.

[25]   Klonoff DC, Buse JB, Nielsen LL, *et al.*, Exenatide effects on diabetes, obesity,cardiovascular risk factors and hepatic biomarkers in patients with type 2 diabetes treated for at least 3 years. Curr Med Res Opin 2008; 24(1): 275-86.

[26]   Arakawa M, Mita T, Azuma K, *et al.* Inhibition of monocyte adhesion to endothelial cells and attenuation of atherosclerotic lesion by a glucagon-like peptide-1 receptor agonist, exendin-4. Diabetes 2010; 59(4): 1030-7.

[27]   Rosenstock J, Klaff LJ, Schwartz S, *et al.* Effects of exenatide and life-style modification on body weight and glucose tolerance in obese subjects with and without pre-diabetes. Diabetes Care 2010; 33: 1173-5.

[28]   Armato J, DeFronzo RA, Abdul-Ghani M, Ruby R. Successful treatment of prediabetes in clinical practice: targeting insulin resistance and β-cell dysfunction. Endocr Pract 2012; 18: 342-50.

[29]   Elkind-Hirsch K, Marrioneaux O, Bhushan M, Vernor D, Bhushan R. Comparison of single and combined treatment with exenatide and metformin on menstrual cyclicity in overweight women with polycystic ovary syndrome. J Clin Endocrinol Metab (Seoul) 2008; 93: 2670-8.

[30]   Koska J, Schwarts EA, Mullin MP, Schwenke DC, Reaven PD. Improvement of postprandial endothelial function af-ter a single dose of exenatide in individuals with impaired glucose tolerance and recent-onset type 2 diabetes. Diabetes Care 2010; 33: 1028-30.

[31]   Schwartz EA, Koska J, Mullin MP, Syoufi I, Schwenke DC, Reaven PD. Exenatide suppresses postprandial elevations in lipids and lipoproteins in individuals with impaired glucose tolerance and recent onset type 2 diabetes mellitus. Atherosclerosis 2010; 212: 217-22.

[32]   Kelly AS, Bergenstal RM, Gonzalez-Campoy JM, Katz H, Bank AJ. Effects of exenatide *vs.* metformin on endothelial function in obese patients with pre-diabetes: a randomized trial. Cardiovasc Diabetol 2012; 11: 64.

[33]   Rosenstock J, Shenouda SK, Bergenstal RM, *et al.* Baseline factors associated with glycemic control and weight loss when exenatide twice daily is added to optimized insulin glargine in patients with type 2 diabetes. Diabetes Care 2012 May; 35(5): 955-8.

[34]   Shao N, Kuang HY, Hao M, Gao XY, Lin WJ, Zou W. Benefits of exenatide on obesity and non-alcoholic fatty liver disease with elevated liver enzymes in patients with type 2 diabetes. Diabetes Metab Res Rev 2014; Sep; 30(6): 521-9.

[35]   Woo JS, Weon K, Ha SJ, *et al.* Cardioprotective effects of exenatide in patients with ST-segment-elevation myocardial infarction undergoing primary percutaneous coronary intervention: results of exenatide myocardial protection in revascularization study. Arterioscler Thromb Vasc Biol 2013; Sep; 33(9): 2252-60.

[36]   Lønborg J, Vejlstrup N, Kelbæk H, *et al.* Exenatide reduces reperfusion injury in patients with ST-segment elevation myocardial infarction. Eur Heart J 2012; 33(12): 1491-9.

[37]   Lønborg J, Kelbæk H, Vejlstrup N, *et al.* Exenatide reduces final infarct size in patients with ST-segment-elevation myocardial infarction and short-duration of ischemia. Circ Cardiovasc Interv 2012; Apr; 5(2): 288-95.

[38]   Agersø H, Jensen LB, Elbrønd B, Rolan P, Zdravkovic M. The pharmacokinetics, pharmacodynamics, safety and toler-ability of NN2211, a new long-acting GLP-1 derivative, in healthy men. Diabetologia 2002; 45: 195-202.

[39]   Bode B. An overview of the pharmacokinetics, efficacy and safety of liraglutide. Diabetes Res Clin Pract 2012; 97: 27-42.

[40] Astrup A, Rössner S, Van Gaal L, *et al*. Effects of liraglutide in the treatment of obesity: a randomised, double-blind, placebo-controlled study. Lancet 2009; 374: 1606-16

[41] Astrup A, Carraro R, Finer N, *et al*. Safety, tolerability and sustained weight loss over 2 years with the once-daily human GLP-1 analog, liraglutide. Int J Obes (Lond) 2012; 36: 843-54.

[42] Pi-Sunyer X, Astrup A, Fujioka K, *et al*.SCALE obesity and prediabetes NN8022-1839 study group. A randomized, controlled trial of 3.0 mg of liraglutide in weight management. N Engl J Med 2015; Jul 2; 373(1): 11-22.

[43] Kim SH, Abbasi F, Lamendola C, *et al*. Benefits of liraglutide treatment in overweight and obese older individuals with prediabetes. Diabetes Care 2013; 36: 3276-82.

[44] Armstrong MJ, Houlihan DD, Rowe IA, *et al*. Safety and efficacy of liraglutide in patients with type 2 diabetes and elevated liver enzymes: individual patient data meta-analysis of the LEAD program. Aliment Pharmacol Ther 2013; Jan; 37(2): 234-42.

[45] Armstrong MJ, Gaunt P, Aithal GP, *et al*. Liraglutide safety and efficacy in patients with non-alcoholic steatohepatitis (LEAN): a multicentre, double-blind, randomised, placebo-controlled phase 2 study. Lancet 2015; Nov 19 [Epub ahead of print].

[46] Daniels GH, Hegedüs L, Marso SP, *et al*.LEADER Trial Investigators. LEADER 2: baseline calcitonin in 9340 people with type 2 diabetes enrolled in the liraglutide effect and action in diabetes: Evaluation of cardiovascular outcome Results (LEADER) trial: preliminary observations. Diabetes Obes Metab 2015; May; 17(5): 477-86.

[47] Morales J, Merker L. Minimizing hypoglycemia and weight gain with intensive glucose control: Potential benefits of a new combination therapy (IDegLira). Adv Ther 2015; May; 32(5): 391-403.

[48] Lau J, Bloch P, Schäffer L, *et al*. Discovery of the once-weekly glucagon-like peptide-1 (GLP-1) analogue semaglutide. J Med Chem 2015; Sep 24; 58(18): 7370-80.

[49] Nauck MA, Petrie JR, Sesti G, *et al*. A phase 2, randomized, dose-finding study of the novel once-weekly human GLP-1 analog, semaglutide, compared with placebo and open-label liraglutide in patients with type 2 diabetes. Diabetes Care 2015; Sep 10 [Epub ahead of print].

[50] Mikkel Christensen, Filip K Knop TVJJH. Lixisenatide for type 2 diabetes mellitus. Expert Opin Investig Drugs 2011; 20(4): 549-57.

[51] Kapitza C, Forst T, Coester HV, Poitiers F, Ruus P, Hincelin-Méry A. Pharmacodynamic characteristics of lixisenatide once daily versus liraglutide once daily in patients with type 2 diabetes insufficiently controlled on metformin. Diabetes Obes Metab 2013; Jul; 15(7): 642-9.

[52] Pfeffer MA, Claggett B, Diaz R, *et al*. Lixisenatide in patients with type 2 diabetes and acute coronary syndrome. N Engl J Med 2015; Dec 3; 373(23): 2247-57.

[53] Brønden A, Naver SV, Knop FK, Christensen M. Albiglutide for treating type 2 diabetes: an evaluation of pharmacokinetics/pharmacodynamics and clinical efficacy. Expert Opin Drug Metab Toxicol 2015; 11(9): 1493-503.

[54] Fisher M, Petrie MC, Ambery PD, Donaldson J, Ye J, McMurray JJ. Cardiovascular safety of albiglutide in the Harmony programme: a meta-analysis. Lancet Diabetes Endocrinol 2015; Sep; 3(9): 697-703.

[55] Sanford M. Dulaglutide: first global approval. Drugs 2014; Nov; 74(17): 2097-103.

[56] Creutzfeldt W. The incretin concept today. Diabetologia 1979; 16: 75-85.

[57]   Crepaldi G, Carruba M, Comaschi M, Del Prato S, Frajese G, Paolisso G. Dipeptidyl peptidase 4 (DPP-4) inhibitors and their role in Type 2 diabetes management. J Endocrinol Invest 2007; 30: 610-4.

[58]   Farilla L, Bulotta A, Hirshberg B, *et al.* Glucagon-like peptide 1 inhibits cell apoptosis and improves glucose responsiveness of freshly isolated human islets. Endocrinology 2003; 144 5149-58.

[59]   Tricco AC, Antony J, Khan PA, *et al.* Safety and effectiveness of dipeptidyl peptidase-4 inhibitors versus intermediate-acting insulin or placebo for patients with type 2 diabetes failing two oral antihyperglycaemic agents: a systematic review and network meta-analysis. BMJ Open 2014 Dec 23; 4(12): e005752.

[60]   Scheen AJ. DPP-4 inhibitors in the management of type 2 diabetes: a critical review of head-to-head trials. Diabetes Metab 2012; 38: 89-101.

[61]   Schernthaner G, Mogensen CE, Schernthaner GH. The effects of GLP-1 analogues, DPP-4 inhibitors and SGLT2 inhibitors on the renal system. Diab Vasc Dis Res 2014 Sep; 11(5): 306-23.

[62]   Dhillon S. Sitagliptin: a review of its use in the management of type 2 diabetes mellitus. Drugs 2010; 70: 489-512

[63]   Bock G, Dalla Man C, Micheletto F, *et al.* The effect of DPP-4 inhibition with sitagliptin on incretin secretion and on fasting and postprandial glu-cose turnover in subjects with impaired fasting glucose. Clin Endocrinol (Oxf) 2010; 73: 189-96.

[64]   Perreault L, Man CD, Hunerdosse DM, Cobelli C, Bergman BC. Incretin action maintains insulin secretion, but not he-patic insulin action, in people with impaired fasting glucose. Diabetes Res Clin Pract 2010; 90: 87-94.

[65]   Hage C, Brismar K, Efendic S, Lundman P, Rydén L, Mellbin L. Sitagliptin improves beta-cell function in patients with acute coronary syndromes and newly diagnosed glucose abnormalities--the BEGAMI study. J Intern Med 2013; 273: 410-21.

[66]   Richter B, Bandeira-Echtler R, Bergerhoff K, Lerch CL. Dipeptidyl peptidase 4 (DPP4) inhibitors for Type 2 diabetes mellitus (review). Cochrane Database of Systematic Reviews issue 2008; 2 Art No.: CD006739.

[67]   Gadsby R. Efficacy and safety of sitagliptin in the treatment of type 2 diabetes. Clin Med Ther 2009; 1: 53-62.

[68]   Fukuhara T, Hyogo H, Ochi H, *et al.* Efficacy and safety of sitagliptin for the treatment of nonalcoholic fatty liver disease with type 2 diabetes mellitus. Hepatogastroenterology 2014; Mar-Apr; 61(130): 323-8.

[69]   Horie A, Tokuyama Y, Ishizuka T, *et al.* The dipeptidyl peptidase-4 inhibitor vildagliptin has the capacity to repair β-cell dysfunction and insulin resistance. Horm Metab Res 2014; Oct; 46(11): 814-8.

[70]   Noguchi K, Hirota M, Miyoshi T, Tani Y, Noda Y, Ito H, Nanba S. Single administration of vildagliptin attenuates postprandial hypertriglyceridemia and endothelial dysfunction in normoglycemic individuals. Exp Ther Med 2015; Jan; 9(1): 84-8.

[71]   Okamoto A, Yokokawa H, Sanada H. Dipeptidyl peptidase-4 inhibitor (vildagliptin) improves glycemic control after meal tolerance test by suppressing glucagon release. Drugs R D 2014; Dec; 14(4): 227-32.

[72]   Utzschneider KM, Tong J, Montgomery B, *et al.* The dipeptidyl peptidase-4 inhibitor vildagliptin improves beta-cell func-tion and insulin sensitivity in subjects with impaired fasting glucose. Diabetes Care 2008; 31: 108-13.

[73] Rosenstock J, Foley JE, Rendell M, *et al*. Effects of the dipeptidyl peptidase IV inhibitor vildagliptin on incretin hormones, islet function, and postprandial glycemia in subjects with impaired glucose tolerance. Diabetes Care 2008; 31: 30-5.

[74] Werzowa J, Hecking M, Haidinger M, *et al*. Vildagliptin and pioglitazone in patients with impaired glucose tolerance after kidney transplantation: a randomized, placebo-controlled clinical trial. Transplantation 2013; 95: 456-62.

[75] Mari A, Sallas WM, He YL, *et al*. Vildagliptin, a dipeptidyl peptidase-IV inhibitor, improves model-assessed beta-cell function in patients with type 2 diabetes. J Clin Endocrinol Metab 2005; 90: 4888-94.

[76] Saab C, Al-Saber FA, Haddad J, Jallo MK, Steitieh H, Bader G, Ibrahim M. Effectiveness and tolerability of second-line treatment with vildagliptin versus other oral drugs for type 2 diabetes in a real-world setting in the Middle East: results from the EDGE study. Vasc Health Risk Manag 2015; Feb 24; 11: 149-55.

[77] Kamenov Z. Effectiveness and tolerability of second-line therapy with vildagliptin versus other oral agents in type 2 diabetes (EDGE): Post Hoc Sub-Analysis of Bulgarian Data. Diabetes Ther 2014; Dec; 5(2): 483-98.

[78] Dejager S, Schweizer A, Foley JE. Evidence to support the use of vildagliptin monotherapy in the treatment of type 2 diabetes mellitus. Vasc Health Risk Manag 2012; 8: 339-48.

[79] Pan C, Yang W, Barona JP,*et al*. Comparison of vildagliptin and acarbose monotherapy in patients with type 2 diabetes: a 24-week, double-blind, randomized trial. Diabet Med 2008; 25: 435-41.

[80] Iwamoto Y, KashiwagiA, Yamada N,*et al*. Efficacy and safety of vildagliptin and voglibose in Japanese patients with type 2 diabetes: a 12-week, randomized, double-blind, active-controlled study. Diabetes Obes Metab 2010; 12: 700-8.

[81] Sicras-Mainar A, Navarro-Artieda R. Use of metformin and vildagliptin for treatment of type 2 diabetes in the elderly. Drug Des Devel Ther 2014; Jun 18; 8: 811-8.

[82] Kudo-Fujimaki K, Hirose T, Yoshihara T, *et al*. Efficacy and safety of nateglinide plus vildagliptin combination therapy compared with switching to vildagliptin in type 2 diabetes patients inadequately controlled with nateglinide. J Diabetes Investig 2014; Jul; 5(4): 400-9.

[83] Su Y, Su YL, Lv LF, Wang LM, Li QZ, Zhao ZG. Randomized controlled clinical trial of a combination therapy of vildagliptin plus an α-glucosidase inhibitor for patients with type II diabetes mellitus. Exp Ther Med 2014; Jun; 7(6): 1752-6.

[84] Viriato D, Calado F, Gruenberger JB, *et al*. Cost-effectiveness of metformin plus vildagliptin compared with metformin plus sulphonylurea for the treatment of patients with type 2 diabetes mellitus: a Portuguese healthcare system perspective. J Med Econ 2014; Jul; 17(7): 499-507.

[85] Leibowitz G, Cahn A, Bhatt DL, *et al*. Impact of treatment with saxagliptin on glycaemic stability and β-cell function in the SAVOR-TIMI 53 study. Diabetes Obes Metab 2015; May; 17(5): 487-94.

[86] Neumiller JJ. Efficacy and safety of saxagliptin as add-on therapy in type 2 diabetes. Clin Diabetes 2014; Oct; 32(4): 170-7.

[87] Toth PP. Overview of saxagliptin efficacy and safety in patients with type 2 diabetes and cardiovascular disease or risk factors for cardiovascular disease. Vasc Health Risk Manag 2014; Dec 23; 11: 9-23.

[88] Bryzinski B, Allen E, Cook W, Hirshberg B. Saxagliptin efficacy and safety in patients with type 2 diabetes receiving concomitant statin therapy. J Diabetes Complications 2014; Nov-Dec; 28(6): 887-93.

[89]   Rosenstock J, Sankoh S, List JF. Glucose-lowering activity of the dipeptidyl peptidase-4 inhibitor saxagliptin in drug-naive patients with type 2 diabetes. Diabetes Obes Metab 2008; 10: 376-86.

[90]   Rosenstock J Aguilar-Salinas C, Klein E,*et al.* Effect of saxaglipin monotherapy in treatment-naive patients with type 2 diabetes. Curr Med Res Opin 2009; 25: 2401-11.

[91]   Jadzinsky M, Pfützner A, Paz-Pacheco E, *et al.* Saxagliptin given in combination with metformin as initial therapy improves glycemic control in patients with type 2 diabetes compared with either monotherapy: a randomized controlled trial. Diabetes Obes Metab 2009; 11: 611-22.

[92]   Wang MM, Lin S, Chen YM, *et al.*, Saxagliptin is similar in glycaemic variability more effective in metabolic control than acarbose in aged type 2 diabetes inadequately controlled with metformin. Diabetes Res Clin Pract 2015; Jun; 108(3): e67-70.

[93]   Gu S, Deng J, Shi L, Mu Y, Dong H. Cost-effectiveness of saxagliptin versus glimepiride as a second-line therapy added to metformin in Type 2 diabetes in China. J Med Econ 2015; May 7: 1-40.

[94]   Schernthaner G, Durán-Garcia S, Hanefeld M, *et al.*, Efficacy and tolerability of saxagliptin compared with glimepiride in elderly patients with type 2 diabetes: a randomized, controlled study (GENERATION). Diabetes Obes Metab 2015; Mar 12.

[95]   Iqbal N, Allen E, Öhman P. Long-term safety and tolerability of saxagliptin add-on therapy in older patients (aged ≥ 65 years) with type 2 diabetes. Clin Interv Aging 2014; Sep 4; 9: 1479-87.

[96]   Konya H, Yano Y, Matsutani S, *et al.*, Profile of saxagliptin in the treatment of type 2 diabetes: focus on Japanese patients. Ther Clin Risk Manag 2014; Jul 11; 10: 547-58.

[97]   Friedrich C, Jungnik A, Retlich S, Ring A, Meinicke T. Bioequivalence of Linagliptin 5 mg once daily and 2.5 mg twice daily: pharmacokinetics and pharmacodynamics in an open-label crossover trial. Drug Res (Stuttg) 2014; May; 64(5): 269-75.

[98]   Heise T, Graefe-Mody EU, Hüttner S, Ring A, Trommeshauser D, Dugi KA. Pharmacokinetics, pharmacodynamics and tolerability of multiple oral doses of linagliptin, a dipeptidyl peptidase-4 inhibitor in male type 2 diabetes patients. Diabetes Obes Metab 2009 Aug; 11(8): 786-94.

[99]   Del Prato S, Barnett AH, Huisman H, Neubacher D, Woerle HJ, Dugi KA. Effect of linagliptin monotherapy on glycaemic control and markers of β-cell function in patients with inadequately controlled type 2 diabetes: a randomized controlled trial. Diabetes Obes Metab 2011; Mar; 13(3): 258-67.

[100]   Kawamori R, Inagaki N, Araki E, *et al.* Linagliptin monotherapy provides superior glycaemic control versus placebo or voglibose with comparable safety in Japanese patients with type 2 diabetes: a randomized, placebo and active comparator-controlled, double-blind study. Diabetes Obes Metab 2012; 14: 348-57.

[101]   Gallwitz B, Rosenstock J, Emser A, von Eynatten M, Woerle HJ. Linagliptin is more effective than glimepiride at achieving a composite outcome of target HbA$_1$c < 7% with no hypoglycaemia and no weight gain over 2 years. Int J Clin Pract 2013; Apr; 67(4): 317-21.

[102]   McGill JB, Barnett AH, Lewin AJ, *et al.* Linagliptin added to sulphonylurea in uncontrolled type 2 diabetes patients with moderate-to-severe renal impairment. Diab Vasc Dis Res 2014; Jan; 11(1): 34-40.

[103]   McGill JB, Sloan L, Newman J, *et al.* Long-term efficacy and safety of linagliptin in patients with type 2 diabetes and severe renal impairment: a 1-year, randomized, double-blind, placebo-controlled study. Diabetes Care 2013; Feb; 36(2): 237-44.

[104]   Del Prato S, Taskinen MR, Owens DR, *et al.* Efficacy and safety of linagliptin in subjects with type 2 diabetes mellitus and poor glycemic control: pooled analysis of data from three placebo-controlled phase III trials. J Diabetes Complications 2013; May-Jun; 27(3): 274-9.

[105]   Haak T, Meinicke T, Jones R, Weber S, von Eynatten M, Woerle HJ. Initial combination of linagliptin and metformin in patients with type 2 diabetes: efficacy and safety in a randomised, double-blind 1-year extension study. Int J Clin Pract 2013; Dec; 67(12): 1283-93.

[106]   Yki-Järvinen H, Rosenstock J, Durán-Garcia S, *et al.*, Effects of adding linagliptin to basal insulin regimen for inadequately controlled type 2 diabetes: a ≥52-week randomized, double-blind study. Diabetes Care 2013; Dec; 36(12): 3875-81.

[107]   Barnett AH, Huisman H, Jones R, von Eynatten M, Patel S, Woerle HJ. Linagliptin for patients aged 70 years or older with type 2 diabetes inadequately controlled with common antidiabetes treatments: a randomised, double-blind, placebo-controlled trial. Lancet Diabetes Endocrinol 2013; Oct 26; 382(9902): 1413-23.

[108]   Zeng Z, Yang JK, Tong N, *et al.* Efficacy and safety of linagliptin added to metformin and sulphonylurea in Chinese patients with type 2 diabetes: a sub-analysis of data from a randomised clinical trial. Curr Med Res Opin 2013; Aug; 29(8): 921-9.

[109]   Inagaki N, Watada H, Murai M, *et al.* Linagliptin provides effective, well-tolerated add-on therapy to pre-existing oral antidiabetic therapy over 1 year in Japanese patients with type 2 diabetes. Diabetes Obes Metab 2013; Sep; 15(9): 833-43.

[110]   Gomis R, Espadero R, Jones R, Woerle HJ, Dugi KA. Efficacy and safety of initial combination therapy with linagliptin and pioglitazone in patients with inadequately controlled type 2 diabetes: a randomized, double-blind, placebo-controlled study. Diabetes Obes Metab 2011; Jul; 13(7): 653-61.

[111]   Friedrich C, Shi X, Zeng P, Ring A, Woerle HJ, Patel S. Pharmacokinetics of single and multiple oral doses of 5 mg linagliptin in healthy Chinese volunteers. Int J Clin Pharmacol Ther 2012; Dec; 50(12): 889-95.

[112]   Seino Y, Yabe D. Alogliptin benzoate for the treatment of type 2 diabetes. Expert Opin Pharmacother 2014; 15: 851-63.

[113]   DeFronzo RA, Fleck PR, Wilson CA, Mekki Q; Alogliptin Study 010 Group. Efficacy and safety of the dipeptidyl peptidase-4 inhibitor alogliptin in patients with type 2 diabetes and inadequate glycemic control: a randomized, double-blind, placebo-controlled study. Diabetes Care 2008; Dec; 31(12): 2315-7.

[114]   Berhan A, Berhan Y. Efficacy of alogliptin in type 2 diabetes treatment: a meta-analysis of randomized double-blind controlled studies. BMC Endocr Disord 2013; 13: 9.

[115]   Seino Y, Fujita T, Hiroi S, Hirayama M, Kaku K. Efficacy and safety of alogliptin in Japanese patients with type 2 diabetes mellitus: a randomized, double-blind, dose-ranging comparison with placebo, followed by a long-term extension. Curr Med Res Opin 2011; 27: 1781-92.

[116]   Kutoh E, Ukai Y. Alogliptin as an initial therapy in patients with newly diagnosed, drug-naive type 2 diabetes: a randomized, control trial. Endocrine 2012; 41: 435-41.

[117] Rosenstock J, Inzucchi SE, Seufert J, Fleck PR, Wilson CA, Mekki Q. Initial combination therapy with alogliptin and pioglitazone in drug-naïve patients with type 2 diabetes. Diabetes Care 2010; Nov; 33(11): 2406-8.

[118] Pratley RE, Reusch JE, Fleck PR, Wilson CA, Mekki Q; Alogliptin Study 009 Group. Efficacy and safety of the dipeptidyl peptidase-4 inhibitor alogliptin added to pioglitazone in patients with type 2 diabetes: a randomized, double-blind, placebo-controlled study. Curr Med Res Opin 2009 Oct; 25(10): 2361-71.

[119] Nauck MA, Ellis GC, Fleck PR, Wilson CA, Mekki Q; Alogliptin Study 008 Group. Efficacy and safety of adding the dipeptidyl peptidase-4 inhibitor alogliptin to metformin therapy in patients with type 2 diabetes inadequately controlled with metformin monotherapy: a multicentre, randomised, double-blind, placebo-controlled study. Int J Clin Pract 2009; Jan; 63(1): 46-55.

[120] Bosi E, Ellis GC, Wilson CA, Fleck PR. Alogliptin as a third oral antidiabetic drug in patients with type 2 diabetes and inadequate glycaemic control on metformin and pioglitazone: a 52-week, randomized, double-blind, active-controlled, parallel-group study. Diabetes Obes Metab 2011; Dec; 13(12): 1088-96.

[121] Pratley RE, Kipnes MS, Fleck PR, Wilson C, Mekki Q; Alogliptin Study 007 Group. Efficacy and safety of the dipeptidyl peptidase-4 inhibitor alogliptin in patients with type 2 diabetes inadequately controlled by glyburide monotherapy. Diabetes Obes Metab 2009; Feb; 11(2): 167-76.

[122] Rosenstock J, Rendell MS, Gross JL, Fleck PR, Wilson CA, Mekki Q. Alogliptin added to insulin therapy in patients with type 2 diabetes reduces HbA(1C) without causing weight gain or increased hypoglycaemia. Diabetes Obes Metab 2009; Dec; 11(12): 1145-52.

[123] Del Prato S, Camisasca R, Wilson C, Fleck P. Durability of the efficacy and safety of alogliptin compared with glipizide in type 2 diabetes mellitus: a 2-year study. Diabetes Obes Metab 2014; Dec; 16(12): 1239-46.

[124] White WB, Cannon CP, Heller SR, *et al.*, Alogliptin after acute coronary syndrome in patients with type 2 diabetes. N Engl J Med 2013; Oct 3 ; 369(14): 1327-35.

[125] C.E. Ezenwaka, O.Okoye, C. Esonwune, *et al.* High Prevalence of Abdominal Obesity Increases the Risk of the Metabolic Syndrome in Nigerian Type 2 Diabetes Patients: Using the International Diabetes Federation Worldwide Definition. Metab Syndr Relat Disord 2014; 12: 277-82.

[126] J.A.Beckman, M.A.Creager, P. Libby. Diabetes and atherosclerosis: epidemiology, pathophysiology, and management. JAMA internal medicine 2002; 287: 2570-81.

[127] P.W. Wilson, R.B.D'Agostino, H. Parise, L. Sullivan, J.B.Meigs. Metabolic syndrome as a precursor of cardiovascular disease and type 2 diabetes mellitus. Circulation 2005; 112: 3066-72.

[128] H. Nagao, S.Kashine, H. Nishizawa, *et al.*, Vascular complications and changes in body mass index in Japanese type 2 diabetic patients with abdominal obesity. Cardiovasc Diabetol 2013; 12: 88.

[129] Scheen AJ, Radermecker RP. Addition of incretin therapy to metformin in type 2 diabetes. Lancet 2010; 375: 1410-12.

[130] Sun F, Wu S, Guo S, *et al.*, Effect of GLP-1 receptor agonists on waist circumference among type 2 diabetes patients: a systematic review and network meta-analysis. Endocrine 2015; Apr; 48(3): 794-803.

[131] Garber AJ. Long-acting glucagon-like peptide 1 receptor agonists: a review of their efficacy and tolerability. Diabetes Care 2011; 34 Suppl 2: S279-S84

[132] Dicker D. DPP-4 inhibitors: impact on glycemic control and cardiovascular risk factors. Diabetes Care 2011; 34 Suppl 2: S276-S8

[133] Gooßen K, Gräber S. Longer term safety of dipeptidyl peptidase-4 inhibitors in patients with type 2 diabetes mellitus: systematic review and meta-analysis. Diabetes Obes Metab 2012; 14: 1061-72.

[134] ACCORD Study Group. Gerstein HC MM, Genuth S, *et al.* Long-term effects of intensive glucose lowering on cardiovascular outcomes. N Engl J Med 2011; 364(9): 818-28.

[135] Farngren J, Persson M, Schweizer A, *et al.*, Vildagliptin reduces glucagon during hyperglycemia and sustains glucagon counterregulation during hypoglycemia in type 1 diabetes. Clin Endocrinol Metab 2012; 97(10): 3799-806.

[136] Tseng CH, Lee KY, Tseng FH. An updated review on cancer risk associated with incretin mimetics and enhancers. J Environ Sci Health C Environ Carcinog Ecotoxicol Rev 2015; 33: 67-124.

[137] Jung MJ, Kwon SK. Expression of glucagon-like Peptide-1 receptor in papillary thyroid carcinoma and its clinicopathologic significance. Endocrinol Metab (Seoul) 2014; Dec 29; 29(4): 536-44.

[138] Bjerre Knudsen L, Madsen LW, Andersen S, *et al.*, Glucagon-like Peptide-1 receptor agonists activate rodent thyroid C-cells causing calcitonin release and C-cell proliferation. Endocrinology 2010; 151: 1473-86.

[139] Elashoff M, Matveyenko AV, Gier B, Elashoff R, Butler PC. Pancreatitis, pancreatic, and thyroid cancer with glucagon-like peptide-1-based therapies Gastroenterology 2011; 141: 150-6.

[140] Rosol TJ. On-target effects of GLP-1 receptor agonists on thyroid C-cells in rats and mice. Toxicol Pathol 2013; 41: 303-9

[141] Hegedüs L, Moses AC, Zdravkovic M, Le Thi T, Daniels GH. GLP-1 and calcitonin concentration in humans: lack of evidence of calcitonin release from sequential screening in over 5000 subjects with type 2 diabetes or nondiabetic obese subjects treated with the human GLP-1 analog, liraglutide. J Clin Endocrinol Metab 2011; 96: 853-60.

[142] Monami M, Ahrén B, Dicembrini I, Mannucci E. Dipeptidyl peptidase-4 inhibitors and cardiovascular risk: a meta-analy-sis of randomized clinical trials. Diabetes Obes Metab 2013; 15: 112-20

[143] Wu D, Li L, Liu C. Efficacy and safety of dipeptidyl peptidase-4 inhibitors and metformin as initial combination therapy and as monotherapy in patients with type 2 diabetes mellitus: a meta-analysis. Diabetes Obes Metab 2014; 16: 30-7

[144] Monami M, Marchinonni N, Mannucci E. Glucagon-like peptide-1 receptor agonists in type 2 diabetes: a meta-analysis of randomized clinical trials. Eur J Endocrinol 2009; 160: 909-17.

[145] Monami M, Cremasco F, Lamanna C, *et al.* Glucagon-like peptide-1 receptor agonists and cardiovascular events: a me-ta-analysis of randomized clinical trials. Exp Diabetes Res 2011; 2011: 215764

[146] Wu S, Hopper I, Skiba M, *et al.*, Dipeptidyl peptidase-4 inhibitors and cardiovascular outcomes: meta-analysis of randomized clinical trials with 55,141 participants. Cardiovasc Ther 2014; 32: 147-58.

[147] Scheen A. Cardiovascular effects of gliptins. Nat Rev Cardiol 2013; 10: 73-84.

[148] Scheen A. Cardiovascular effects of dipeptidyl peptidase-4 inhibitors: from risk factors to clinical outcomes. Postgrad Med 2013; 125: 7-20.

[149] Scirica BM, Bhatt DL, Braunwald E,*et al.*, Saxagliptin and cardiovascular outcomes in patients with type 2 diabetes mellitus. N Engl J Med 2013; 369: 1317-26.

[150] Green JB, Bethel MA, Armstrong PW, *et al.*, Effect of Sitagliptin on Cardiovascular Outcomes in Type 2 Diabetes. N Engl J Med 2015; Jun 8: [Epub ahead of print].

[151] .Petrie JR. The cardiovascular safety of incretin-based therapies: a review of the evidence. Cardiovasc Diabetol 2013; 12: 130.

[152] Matveyenko AV, Dry S, Cox HI, *et al.*, Beneficial endocrine but adverse exocrine effects of sitagliptin in the human islet amyloid polypeptide transgenic rat model of type 2 diabetes: interactions with metformin. Diabetes 2009; 58: 1604-15.

[153] Meier JJ, Nauck MA. Risk of pancreatitis in patients treated with incretin-based therapies. Diabetologia 2014; 57: 1320-4.

[154] Singh S, Chang HY, Richards TM, *et al.*, Glucagonlike peptide 1-based therapies and risk of hospitalization for acute pancreatitis in type 2 diabetes mellitus: a population-based matched case-control study. JAMA internal medicine 2013; 173: 534-9.

[155] Butler AE, Campbell-Thompson M, Gurlo T, *et al.*, Marked expansion of exocrine and endocrine pancreas with incretin therapy in humans with increased exocrine pancreas dysplasia and the potential for glucagon-producing neuroendocrine tumors. Diabetes 2013; 62: 2595-604.

[156] Bonner-Weir S, In't Veld PA, Weir GC. Reanalysis of study of pancreatic effects of incretin therapy: methodological deficiencies. Diabetes Obes Metab 2014; 16: 661-6.

[157] Li L, Shen J, Bala MM, *et al.*, Incretin treatment and risk of pancreatitis in patients with type 2 diabetes mellitus: systematic review and meta-analysis of randomised and non-randomised studies. BMJ 2014; 348: g2366.

[158] Nauck MA. A critical analysis of the clinical use of incretin-based therapies: the benefits by far outweigh the potential risks. Diabetes Care 2013; 36: 2126-32.

[159] Monami M, Dicembrini I, Mannucci E. Dipeptidyl peptidase-4 inhibitors and pancreatitis risk: a meta-analysis of randomized clinical trials. Diabetes Metab Res Rev 2013; 29(8): 624-30.

[160] White WB, Cannon CP, Heller SR,*et al.*, Alogliptin after acute coronary syndrome in patients with type 2 diabetes. N Engl J Med 2013; 369: 1327-35.

[161] Koehler JA, Baggio LL, Lamont BJ, Ali S, Drucker DJ. Glucagon-like peptide-1 receptor activation modulates pancreatitis-associated gene expression but does not modify the susceptibility to experimental pancreatitis in mice. Diabetes 2009; 58 2148-61

[162] Tatarkiewicz K, Smith PA, Sablan EJ, *et al.*, Exenatide does not evoke pancreatitis and attenuates chemically induced pancreatitis in normal and diabetic rodents. Am J Physiol En-docrinol Metab 2010; 299: E1076-E86.

[163] Mizukami H, Inaba W, Takahashi K, Kamata K, Tsuboi K, Yagihashi S. The effects of dipeptidyl-peptidase-IV inhibi-tor, vildagliptin, on the exocrine pancreas in spontaneously diabetic Goto-Kakizaki rats Pancreas 2013; 42: 786-94.

[164] Garg R, Chen W, Pendergrass M. Acute pancreatitis in type 2 diabetes treated with exenatide or sitagliptin: a retrospective observational pharmacy claims analysis. Diabetes Care 2010; 33: 2349-54.

[165] Dore DD, Seeger JD, Arnold Chan K. Use of a claims-based active drug safety surveillance system to assess the risk of acute pancreatitis with exenatide or sitagliptin compared to metformin or glyburide Curr Med Res Opin 2009; 25: 1019-27

[166] Dore DD, Bloomgren GL, Wenten M, *et al.,* A cohort study of acute pancreatitis in relation to exenatide use. Diabetes Obes Metab 2011; 13: 559-66.

[167] Wenten M, Gaebler JA, Hussein M, *et al.,* Relative risk of acute pancreatitis in initiators of exenatide twice daily compared with other anti-diabetic medication: a follow-up study. Diabet Med 2012; 29: 1412-8.

[168] Engel SS, Round E, Golm GT, Kaufman KD, Goldstein BJ. Safety and tolerability of sitagliptin in type 2 diabetes: pooled analysis of 25 clinical studies. Diabetes Ther 2013; 4: 119-45

[169] Monami M, Dicembrini I, Mannucci E. Dipeptidyl peptidase-4 inhibitors and pancreatitis risk: a meta-analysis of randomized clinical trials Diabetes Obes Metab 2014; 16: 48-56.

[170] Faillie JL, Azoulay L, Patenaude V, Hillaire-Buys D, Suissa S. Incretin based drugs and risk of acute pancreatitis in patients with type 2 diabetes: cohort study. BMJ 2014; 348: g2780.

[171] Romley JA, Goldman DP, Solomon M, McFadden D, Pe-ters AL. Exenatide therapy and the risk of pancreatitis and pancreatic cancer in a privately insured population. Diabetes Technol Ther 2012; 14: 904-11

[172] Giorda CB, Picariello R, Nada E, Tartaglino B, Marafetti L, Costa G, Gnavi R. Incretin therapies and risk of hospital ad-mission for acute pancreatitis in an unselected population of European patients with type 2 diabetes: a case-control study. Lancet Diabetes Endocrinol 2014; 2 111-5

[173] Li L, Shen J, Bala MM, *et al.,* Incretin treatment and risk of pancreatitis in patients with type 2 diabetes mellitus: systematic review and meta-analysis of randomised and non-randomised studies. BMJ 2014; 348: g2366

[174] Gokhale M, Buse JB, Gray CL, Pate V, Marquis MA, Stürmer T. Dipeptidyl-peptidase-4 inhibitors and pancreatic cancer: a cohort study. Diabetes Obes Metab 2014; Dec; 16(12): 1247-56.

[175] Wang T, Wang F, Gou Z, *et al.* Using real-world data to evaluate the association of incretin-based therapies with risk of acute pancreatitis: a meta-analysis of 1,324,515 patients from observational studies. Diabetes Obes Metab 2015; Jan; 17(1): 32-41.

[176] Scheen AJ. Pharmacokinetics in patients with chronic liver disease and hepatic safety of incretin-based therapies for the management of type 2 diabetes mellitus. Clin Pharmacokinet 2014; 53: 773-85.

[177] He YL, Sabo R, Campestrini J, *et al.,* The influence of hepatic impairment on the pharmacokinetics of the dipeptidyl peptidase IV (DPP-4) inhibitor vildagliptin. Eur J Clin Pharmacol 2007; 63(7): 677-86.

CHAPTER 7

New and Promising Drugs for Obesity

Natalia G. Vallianou[1,*], Angelos A. Evangelopoulos[2] and Christos E. Kazazis[3]

[1]Evangelismos General Hospital Athens, Greece; [2]Roche Hellas Diagnostics Athens, Greece and [3]Leicester University, London, UK

Abstract: As the obesity epidemic continues, there is demand for more potent and with less side effects anti-obesity drugs. Serotonin is widely- known to participate in the process of satiety. A novel drug that has been associated with serotonin, has been FDA approved in 2012 for the treatment of obesity and is called lorcaserin. It acts as a selective 5-HT2C agonist on pro-opiomelanocortin neurons, thus inducing alpha- melanocyte stimulating hormone (alpha-MSH) and thereby, decreasing appetite. Bupropion also stimulates pro-opiomelanocortin neurons in the hypothalamus resulting in an anorexigenic effect. The combination of phentermine and extended-release topiramate has recently gained approval for the treatment of obesity by the FDA. Phentermine stimulates the release of noradrenaline, dopamine, and serotonin, in a non-selective manner. Phentermine has already been tried as a short-term therapy for obesity, while topiramate, a drug used for the treatment of epilepsy, has been proposed as an anti-obesity drug, based on remarks of weight loss among patients treated with this anticonvulsant drug. Besides, it has been documented that "glucagon-like peptide 1" (GLP-1), and glucagon, together with oxyntomodulin (OXM) have the potential to produce brown adipose tissue (BAT) thermogenesis by means of a centrally involved sympathetic pathway. As the use of "GLP-1 receptor agonists" for treating type 2 diabetic patients has become widespread, this category of drugs could probably be efficacious in combating obesity in the near future. In fact, exanetide and liraglutide have been recently FDA approved as anti-obesity drugs. Orlistat together with cetilistat are gastrointestinal and pancreatic lipase inhibitors, which result in reduced fat absorption, but cetilistat is suggested to have better tolerability than orlistat. Recent data has reported the existence of a new class of neuropeptides, called orexins or hypocretins, which are produced in the hypothalamus and whose actions are mediated by two types of receptors: OX1R and OX2R. More specifically, the orexinergic neurons have been located exclusively in cells in the lateral, dorsomedial and perifornical areas of the hypothalamus. Despite this highly specific anatomical origin, the orexinergic neurons are projected widely into a number of brainstem, cortical and limbic regions. Besides, targeting peroxisome proliferator activated-receptor gamma, seems to be a promising pathway to control the obesity epidemic in the future. Other novel drugs, which have been explored and seem promising against obesity and therefore are worth mentioning are beloranib, velerenib and tesofensine. Tesofensine possesses inhibitory properties

*Corresponding author Natalia G. Vallianou: Evangelismos General Hospital Athens, Greece; Tel: +30 2294092359; E-mail: natalia.vallianou@hotmail.com

Atta-ur-Rahman (Ed)

regarding serotonin, dopamine and mainly noradrenaline. Velneperit is an oral, small-molecule NPY5 receptor antagonist, which acts centrally and is given once daily. Neuropeptide Y (NPY), an appetite-stimulating substance, has a key -role in homeostasis of energy balance, especially under food deprivation conditions. Velneperit counteracts elevated NPY levels, thus producing sustained weight loss. Beloranib, "an analog of the natural substance fumagillin, is a methionine amino peptidase 2 inhibitor (MetAP2)". The ability of "MetAP2 to suppress extracellular regulated kinases 1 and 2 (ERK1/2)" is thought to act as the major the mechanism associated with beloranib's potential against obesity. Brown adipose tissue possesses too many vessels. It is by means of these precursor cells that are present in vessels, that adipogenesis is controlled by angiogenesis. Adipose-derived precursor cells produce peptides, such as "vascular endothelial growth factor (VEGF)", leptin, resistin, and visfatin, and matrix metalloproteinases (MMPs), which are implicated in and promote angiogenesis. Therefore, new targets such as VEGF, matrix- metalloproteinases and sirtuin receptors are currently being investigated, with regards to their anti-obesity potential.

Keywords: Beloranib, brown adipose tissue, bupropion, cetilistat, GLP-1 analogues, incretin effect, lorcaserin, obesity epidemic, orexins, oxyntomodulin, oxytocin, phentermine/topiramate ER, PPARγ, tesofensine, VEGF, velneperit.

BROWN ADIPOSE TISSUE, OXYNTOMODULIN AND GLP-1 ANALOGUES

Brown adipose tissue possesses too many vessels. It has been documented that between cells of the vessel wall there are precursor cells. It is by means of these precursor cells that adipogenesis is controlled by angiogenesis. Adipose-derived precursor cells produce peptides, such as vascular endothelial growth factor (VEGF), leptin, resistin, and visfatin, and matrix metalloproteinases (MMPs), which are implicated in and promote angiogenesis. Therefore, new targets such as VEGF, matrix- metalloproteinases and sirtuin receptors are currently being investigated, with regards to their anti-obesity potential. The existence of brown adipose tissue (BAT) in newborns that is actually metabolically active has been widely known for a long time [1, 2]. However, it had been thought that metabolically active BAT does not exist in human adults [3, 4].

A remark that has changed this belief came across by means of nuclear medicine [5]. While performing "positron emission tomography (PET-CT Scan)" for cancer staging, a confounding symmetrical uptake of [18F]-fluoro-deoxy-glucose was detected in neck and shoulder areas of some individuals. CT revealed that those areas had features of adipose tissue. The above-mentioned findings led scientists to test the notion, that human adults do have substantial BAT. In April 2009, three independent studies that featured in the "New England Journal of Medicine", actually demonstrated the presence of such tissue in human adults [6-8]. Since then, several studies have shed light on this issue and it is nowadays a well-known fact that, not only infants, but also many adults possess BAT [9-12]. Existing documentation suggests that "adult humans do have BAT, which may be activated". In addition, in obese patients there is a defect in its activation [13]. Several studies have documented that mistakes in the thermogenic properties of BAT, lead to defective thermoregulation as well as in defective body weight control. In human adults, it has been suggested that "activated BAT" may account for "as much as 15 % of energy expenditure" [14]. The intake of excessive dietary calories could induce thermogenesis by BAT. If the organism is unable to use BAT to reduce the effects of large caloric intake, then the result is obesity [15, 16]. Brain is "the major regulator of thermogenesis, by means of sympathetic innervation of BAT" [17]. The physiological regulation of BAT in animals has not been studied extensively [18-20]. In a recent study, researchers have used intracerebroventricular administration of the "proglucagon-derived peptides GLP-1, oxyntomodulin, and glucagon" in aiming to answer this question [21].

OXYNTOMODULIN

Oxyntomodulin (OXM) is peptide consisting of 37 amino acids, secreted in proportion to nutrient ingestion [22-25]. The processing of pro-glucagon is tissue specific, producing different hormones, depending on the specific tissue involved. In pancreatic a cells, pro-hormone convertase 2 (PC2) generates predominantly glucagons, whereas in intestinal "L cells" present in the small intestine, PC 1/3 predominantly produces glicentin, OXM, GLP1, and GLP2 [22-25, 26-32]. Similar processing is considered to occur in the same neurons in the hindbrain. The first evidence of the existence of OXM came to light in 1968, when it was discovered that the gut glucogen like inhibitor consisted of at least two peptides. Two distinct moieties with glucogen like inhibitor were found to be excreted as a response to an oral glucose load [33, 34]. One with a C-terminal octapeptide extension (SP-1, spacer peptide-1, or IP-1, intervening peptide) was named OXM, for its ability to modulate gastric acid secretion in gastric oxyntic glands, and the other with the same C-terminal extension plus an N-terminal consisting of 30 amino acids, was glicentin [35-41]. Larsson *et al.* showed in 1975 that disseminated cells that predominate in the ileum and colon intestinal mucosa store gut-type glucagon, reporting for the first time the spatial tissue distribution of OXM in the gut [42]. The distribution of glucagon- related peptides in the human gastrointestinal tract was later described [43]. Consistent with the properties of a partial agonist, OXM was a functional antagonist of GLP1- induced agonist response in b-arrestin 2 recruitment. It has been suggested that OXM is a GLP1R-biased agonist relative to GLP1, having less preference toward c-AMP signaling relative to phosphorylation of ERK1/2, but similar preference for c-AMP relative to Ca_2C [44]. These findings imply that the GLP1R-mediated *in vivo* 7 effects of OXM might differ from that of GLP1 (45). OXM produces weight loss by suppressing food intake and by increasing "energy expenditure" [46]. OXM is an agonist at GLP1R and GCGR *in vitro* [44, 45, 47]. The anorectic effects of central administrations of GLP1 and OXM evanish by the administration of the GLP1 receptor agonist, exendin (9-39), and are not present in Glp1rK/K rodents, suggesting that the central effect of OXM is mediated by the GLP1R [48-51]. The lack of OXM efficacy seen in Glp1rK/K mice demonstrated that the initial anorectic effect of oxyntomodulin is performed solely *via* activation of GLP1R [47, 51]. However, other properties of OXM are independent of GLP1R, suggesting that OXM has GLP1R- independent as well as dependent effects *in vivo* [47, 51]. These data led several investigators to hypothesize that the differential effect of OXM *versus* GLP1 could be mediated by activation of the GCGR or an as of yet unidentified OXM- specific receptor. This study showed the involvement of the GCGR together with GLP1R activation to the body reducing weight ability of OXM, but it did not completely dismiss the potential contribution of an OXM-specific receptor, as minor weight loss was noticed in those treated with a small molecule GCGR antagonist alone [52]. Repeated administration of glucagons was first shown to inhibit food intake in man over 50 years ago, and aside from its well-known hyperglycemic action, glucagons increases thermogenesis, satiety, lipolysis, fatty acid oxidation, and ketogenesis [53, 54]. A crucial physiological role of glucagon regarding maintenance of whole–body energy homeostasis was demonstrated by a recent study in type 2 diabetic patients, in whom a "dose-dependent increase in body weight was documented", after blockade of the GCGR [54]. Preclinical data support the notion that OXM may have glucose-lowering properties [52-55]. However, the clinical utility of OXM is limited, mainly because of its short circulating half-life [56]. Repeated daily doses of large amounts of peptide would be required to elicit its effect, entailing a treatment regimen inconvenient for patients and not economically viable. Because glucagon and GLP1 share approximately 50 % amino acid sequence identity, several groups have recently developed protease-resistant GLP1R/GCGR dual agonist peptides that are resistant to peptidase degradation [57-59]. Two independent researchers reported that the use of GLP1R/GCGR co-agonists yielded better results, relatively to pure GLP1R agonists in the treatment of rodent obesity, with dual improvement in glycemic control [60]. Two DPP4-resistant OXM analogues have been used to test the effects of "dual agonism regarding the activation of the GLP1R" [61]. "One analogue, being a dual agonist at the GLP1R and GCGR (DualAG), is the oxyntomodulin peptide with a cholesterol group attached to the C-terminal chain". The alternative analogue is a GLP1R-selective analogue (GLPAG), with an equal affinity for the GLP1R, but no much activity at the GCGR, due to a mutation from glutamine to glutamate, which abolished the interaction with the GCGR [62].

Obese rodents administered with DualAG had better weight loss and lipid lowering effect, compared with the GLP1R-selective agonist [63]. Another dual GLP1R and GCGR agonist strategy involved testing several chimeric DPP4-resistant PEGylated peptides. The chimeric peptide was optimized to decrease food intake, reduce body weight, and increase GLP1 activity to neutralize the hyperglycemic effects of glucagon injections to diet-induced obese mice [64]. The enhanced weight loss observed with GLP1R/GCGR dual agonists [65] has triggered important questions about the ideal ratio of receptor activation. In particular, what is the appropriate amount of GLP1R activation, which buffers the hyperglycemic risk posed by GCGR activation. When a long-acting dual agonist was given to Glp1rK/K mice, the decrease in body weight was no longer associated with improvement in glucose metabolism, highlighting the importance of an appropriate GLP1R engagement in preventing GCGR-mediated increase in glucose production. A recent report using receptors with selectivity demonstrated that a dual agonist peptide with comparable functional potencies at the GLP1R and GCGR maximized the weight loss and minimized the hyperglycemic risk associated with GCGR activation in mice [65]. Another OXM analogue, OXM6421, when injected in lean rodents was noticed to exhibit a longer half-life than endogenous oxyntomodulin and also induced a decreased food intake and an enhanced energy expenditure [66]. ZP2929, a chimeric protein able to fully activating both GLP1R and GCGR, ameliorated glycemic control without body weight gain in db/db mice, when combined with long-acting insulin [67]. Recently, Zealand Pharma announced the starting of a Phase I development of ZP2929 for the treatment of T2DM and/or obesity (Diabetes.co.uk, Sep 14, 2012) and Transition Therapeutics confirmed in a press release the completion of a Phase 1 study with TT-401, a weekly GLP1R/GCGR dual agonist developed as an anti-diabetic regimen [www. transitiontherapeutics.com/media/news.php, June 6, 2012].

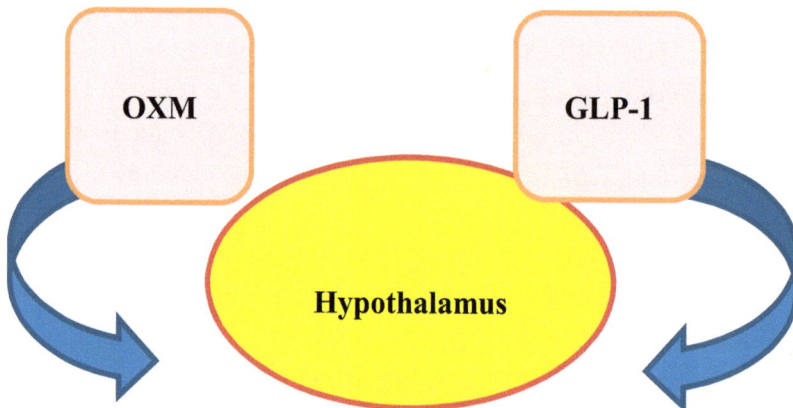

Figure 1: Oxyntomodulin (OXM) and Glucagon-like Peptide -1 (GLP-1) result in different hypothalamic pathways activation, thus suppressing appetite.

GLP-1 ANALOGUES AND THEIR CENTRAL NERVOUS SYSTEM MECHANISMS OF ACTION

It has been suggested that food, especially, calorie-dense, obeso- genic food has been rewarding. Therefore, in order to develop effective anti- obesity treatments, it would be useful to discover the mechanisms, which may decrease the hedonically-driven eating and food reward behavior. In this context, one potentially promising therapeutic agent is glucagon-like- peptide-1 ("GLP-1").

Endogenous GLP-1 is produced in the L-cells in the small intestine and the hindbrain [65-68]. GLP-1 receptors (GLP-1R) may also be stimulated by means of long lasting GLP-1 analogues [69-71]. Several GLP-1 analogues, among them exendin 4 (exanetide) or liraglutide, have gained approval for clinical use in type 2 diabetic patients, to improve glycemic control [72-76]. Much has been learnt about the anatomical, neurochemical, and functional suppressive effects of GLP-1 or its analogues on food intake; GLP-1's ability to suppress food reward behavior is a relatively new concept. Central or peripheral GLP-1 injection has been found to induce decreased food intake in mice and men [71-78]. The literature has primarily focused on hypothalamic and brainstem nuclei, as the key central nervous system targets for anorexic and to some extent gluco-regulatory effects of GLP-1 [79, 80]. Recent data have shown that GLP-1R stimulation reduces food reward behavior [81]. This observation was very important, as it supports the potential activity of GLP-1 in areas classically associated with reward behavior, such as the ventral tegmental areas (VTA) and the nucleus accumbens (NAc). The VTA and its dopaminergic projections to the NAc dominate a goal-directed, motivated role of GLP-1 in reward behavior, to obtain natural reinforcements, like food [82-84]. Many of the classic hormones regulating feeding have a direct impact on the mesolimbic VTA/NAc neurons [85-89]. GLP-1 is the newest member in this group. 13 Unlike many of the gut or fat hormone counterparts (ghrelin or leptin), for which there is little evidence of central production, GLP-1 can be made in the brain (Fig. **2**). Local application of GLP-1 or GLP-1 analogues (*e.g.*, EX4) into the VTA or the NAc alters food motivation/reward. Furthermore, many other GLP-1R expressing central nervous system sites that directly respond to GLP-1, have clear connections to the mesolimbic dopamine circuitry, which plays a key role in food reward behaviors. Also, projections of these hindbrain GLP-1 neurons are widely distributed, giving neuroanatomical support for the diverse physiological and behavioral effects of the endogenous GLP-1. Key mesolimbic reward areas, like the VTA and the NAc, are innervated by the GLP-1 producing neurons, potentially allowing hindbrain GLP-1 to modulate reward behavior directly. Ascending fibers from the caudal nucleus tactus solitarius, identified with anterograde tracing, terminate in the VTA and the NAc [90]. In fact, nearly one third of all the nucleus tactus solitarius GLP-1-producing neurons send ascending fibers to the VTA and the NAc [91-93]. Thus, GLP-1neurons are placed in a very influential position, enabling them to sample the hormonal milieu, visceral sensory and gustatory input, integrate and carry this information directly to the mesolimbic system without any intermediate.

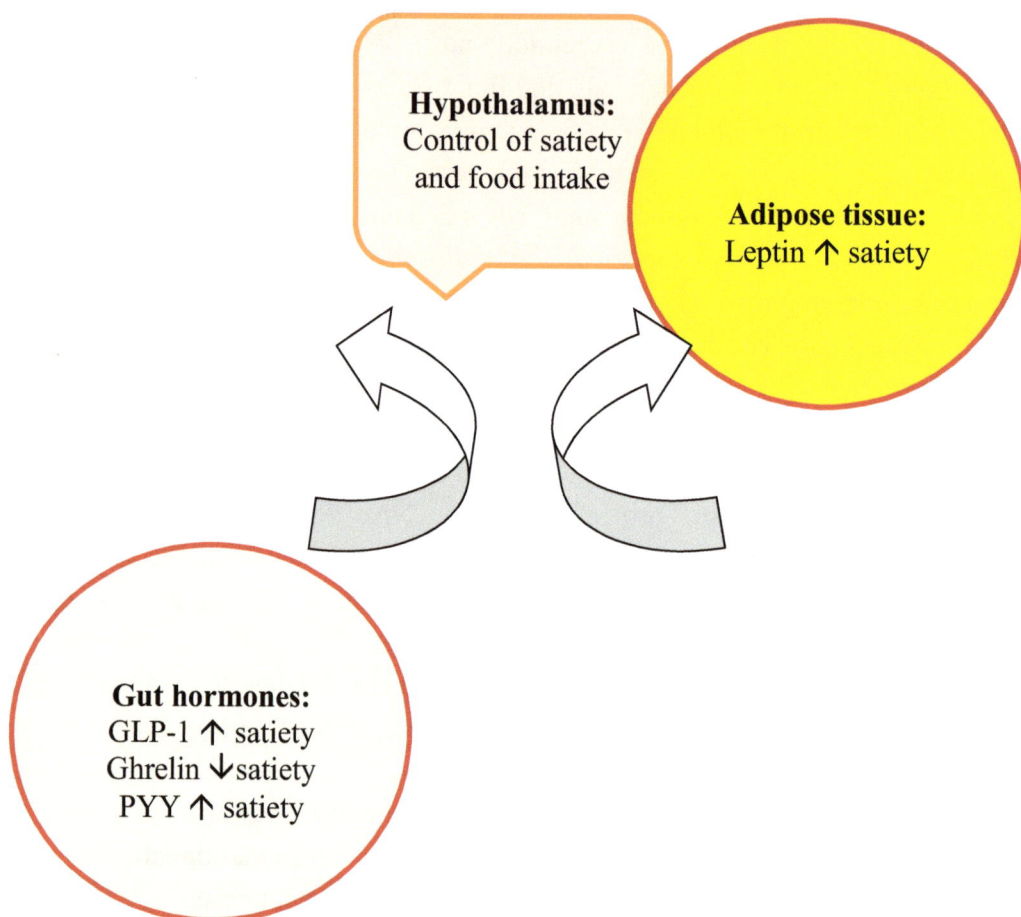

Hypothalamus:
Control of satiety
and food intake

Adipose tissue:
Leptin ↑ satiety

Gut hormones:
GLP-1 ↑ satiety
Ghrelin ↓ satiety
PYY ↑ satiety

Figure 2: Leptin, GLP-1 and PYY increase satiety and decrease food intake, while ghrelin decreases satiety and increases food intake by interacting with the hypothalamic cores.

GLP-1 ANALOGUES AND PERIPHERAL TISSUES MECHANISMS OF ACTION

"Glucagon-like peptide-1 (GLP-1) is a peptide hormone, which stimulates insulin and inhibits glucagon secretion". GLP-1 also reduces appetite and delays gastric emptying, actions resulting in a better glycemic control [94]. The incretin effect, "which accounts for 50-70 % of total insulin secretion after oral glucose load, is the difference in insulin secretory response from oral glucose administration, compared with intravenous glucose administration" [95]. "There are two naturally occurring incretin hormones, which play a major role in the maintenance of glycemic control: glucose dependent insulinotropic polypeptide and GLP- 1. Both have a short half-life because of their rapid inactivation by DPP-4" [96]. "In type 2 diabetic patients, the incretin effect is reduced or even absent" [97, 98]. However, it has been documented that "after administration of GLP-1, the insulin secretory function can be restored in this population", and thus GLP-1 has become an important researching tool for type 2 diabetes mellitus [99]. Exanetide, which is the synthetic form of exendin-4, is an injectable GLP-1 receptor agonist that has been FDA approved for the treatment of type 2 diabetes mellitus [100]. Liraglutide is a GLP-1 analogue, which is also FDA approved for treating type 2 diabetic patients [101]. Liragutide has just been received 14: 1 votes in its favor, by the FDA advisory panel for treating obesity. The specific proposed indication was "for chronic weight management in individuals with a 16 body mass index of 30 kg/m^2 or greater, or 27 kg/m^2 or greater in the presence of at least 1 weight-related comorbidity". Lisixenatide is EMA approved by not yet FDA approved. Other GLP-1 receptor analogues, that are currently in development include taspoglutide and albiglutide, which are long acting analogues, administered once weekly [102, 103]. Gastrointestinal side effects, such as nausea, vomiting and diarrhea, are the most frequently reported adverse side effects seen with this category of drugs. These adverse side effects "usually occur early on the course of treatment, but tend to be transient and rarely cause patient withdrawal from therapy" [104-110]. Cases of pancreatitis have been noticed among patients who have been treated with GLP-1 receptor analogues or DPP-4 inhibitors [111]. No cases of pancreatitis were reported during the DURATION-2 trial [112]. "Large studies, involving diabetic rodents, have failed to show a relationship between GLP-1 agonists, such as exenatide and liraglutide, as well as the DPP-4 inhibitor sitagliptin and pancreatitis" [113, 114]. Until now, large cohort studies have not shown any relationship between incretin-based therapies and pancreatitis [115, 116]. A recently published large study focused on the frequency of acute pancreatitis in diabetic subjects treated with exenatide, sitagliptin or other antidiabetic regimens. "The risk of pancreatitis was high in patients with diabetes compared to patients without diabetes, but there was no increased risk of pancreatitis seen in patients treated with exenatide or sitagliptin, compared to patients who received other diabetic drugs" [117]. The available data "do not support a relationship between incretin therapies and pancreatitis yet". Further studies are needed to clear this issue. Long term exposure to liraglutide has been documented to be related to "thyroid C cell hyperplasia" in mice [118]. On the contrary, "monkeys and humans have lower levels of GLP-1R expression, and administration of liraglutide at high doses has not been shown to produce C-cell proliferation in monkeys". Data from other studies, such as the 1860-Lira- DPP-4 trial, have failed to yield any increase in the mean calcitonin level in patients on liraglutide treatment [119]. Nevertheless, liraglutide is contraindicated in patients with "history of medullary carcinoma or multiple endocrine neoplasia syndrome type 2". Also, rare cases of "papillary thyroid carcinoma" have been identified in patients on liraglutide treatment, an issue that requires further investigation [120]. With these initial promising results from incretin mimetics and incretin enhancers, the next generation of diabetes drugs will likely focus on the alternate delivery for injectables and the combined activation of more than one receptor [121-123]. Simultaneous activation of GLP1R and GCGR, with chimeric peptides, and in the future non-peptide, orally available GLP1R/GCGR dual agonists is a conceivable option to achieve improved therapeutic goals. It will be critical to deepen our knowledge regarding the mechanism of action and how structurally related peptides like GLP1 and glucagon interact with their receptors, respectively. Understanding of receptor oligomerization, heteromerization, and binding co-operativity will allow an improved understanding of how ligands should be designed to maximize the simultaneous activation of these complexes. [124-126]. Finally, characterization of the post-receptor signaling of these closely related GPCR in the glucagon family will allow a better understanding of the pathways that need to be selectively modulated to achieve the desired effect, while avoiding various undesirable adverse effects [127].

TOPIRAMATE

Topiramate is an anticonvulsant drug, which has been approved in the mid-1990s for combating refractory seizures (recommended dose 400 mg/day) in conjunction with other anticonvulsants. It has also been administered for the prevention of migraine (recommended dose 100 mg/day) [128].

Topiramate is a "monosaccharide derivative of the sugar d- fructose" [129]. It has been suggested that topiramate acts "on kainate/alpha-amino-3-hydroxy-5-methylisoxozole-4-propionic acid glutamate receptors" [130, 131]. While testing topiramate for treatment of mood disorders, it was noticed that the agent might decrease the weight gain associated with anti-depressant treatment, and an interesting study has established that these effects have been dose-dependent [132]. This explains the reason, why some physicians have tested topiramate "off label" for weight reduction, for which it had no regulatory approval. The mechanisms of action for topiramate's weight loss effect have not been fully elucidated. Topiramate can reduce food intake by means of carbonic-anhydrase inhibition on taste.

Apart from that, its effects on GABA transmission and the involvement between "GABA and leptin pathways" are known to be implicated in appetite [133, 134]. Topiramate's antagonism of "a-amino-3-hydroxy-5-methyl-4- isoxazole-propionic acid kainite receptors" may reduce "compulsive or addictive food cravings, which is supported by topiramate's effectiveness in improving binge eating disorder" [135-137]. Topiramate's activation of GABA receptors is likely to "reduce night- time and deprivation-induced feeding, which results in persistent weight loss in combination with decreased feeding" [138].

While animal studies suggest that "topiramate could increase the levels of neuropeptide Y (NPY) in the hypothalamus, topiramate might also increase levels of hypothalamic corticotropic-releasing hormone, which are likely catabolic: [139, 140]. While some studies have suggested that "leptin levels decrease with topiramate-induced weight reduction, other studies indicate that leptin levels are only modestly decreased, which may be another explanation that may support the persistence of weight loss" [141, 142]. Anorexia is common side effect of topiramate. This would also explain the persistence in weight loss. Topiramate may also affect energy expenditure [143, 144]. It could reduce energy storage and its usage, resulting in increased energy expenditure [145]. Topiramate might "inhibit adipocyte mitochondrial carbonic anhydrase isozyme V, thus inhibiting carbonic anhydrase- mediated lipogenesis" [146-148]. Topiramate probably also affects "lipoprotein lipase activity in white and brown adipose tissue, which would limit free fatty acid substrate for lipogenesis and could increase thermogenesis" [149].

PHENTERMINE

Phentermine was approved in 1959 by the FDA, as a short-term (<12 weeks) treatment for weight loss in combination with appropriate nutrition and physical exercise (recommended dose 37.5 mg/ day) [150- 153]. It is available in some countries but not in Europe, where it was withdrawn in 2000 due to an unfavorable risk to benefits ratio [154]. Phentermine is an "atypical amphetamine analogue, which acts to increase norepinephrine in the CNS, thus, suppressing appetite" [155, 156]. Phentermine potently releases norepinephrine, but is effective on the release of dopamine and serotonin, too, to a lesser extent [157]. The drug is a "weak substrate of the serotonin transporter and has minimal effects on plasma serotonin *in vivo*" [158-160]. "Although increased catecholamine release may be thought to increase the odds for adverse cardiovascular effects, norepinephrine released in the brain may act to decrease sympathetic activity, through a "clonidine-like" effect" [161- 163].

Phentermine is the most widely used weight-loss drug in the USA; "phentermine HCl, an immediate release formulation that undergoes rapid dissolution and absorption in the gastrointestinal tract, is currently approved at a dose of 15.0-37.5 mg/day for the short-term (up to 12 weeks) treatment of obesity" [164]. "A phentermine resin that slowly releases active drug into the gastrointestinal system is also approved in the USA for the short-term treatment of obesity". Phentermine (either phentermine HCl 37.5 mg/day or phentermine resin 15-30 mg/day) has been evaluated in several studies for up to 36 weeks, and has accomplished decreases in body weight and waist circumference, and has achieved weight loss of at least 5 or 10 % relative to placebo [165- 167]. In 1968, Munro *et al.* studied weight loss with phentermine (30 mg/day) in a 36-week trial among overweight and obese women, and documented weight loss with both continuous and intermittent administration of phentermine (12.2 and 13.0 kg, respectively) [167]. "Although no published studies describing long term, randomized, controlled weight-loss trial data for phentermine are currently available, Hendricks *et al.* in 2011 observed that patients (n=300) treated with phentermine monotherapy, at doses ranging from 15 to 37.5 mg/day, showed more weight loss for 1 through 104 weeks of treatment (12.7 %), compared with those on a low- carbohydrate ketogenic di*et al*one" (8.4 %; P < 0.0144) [168]. The notion that phentermine increases heart rate and BP may be due to the assumption of "amphetamine-like" adverse side effects. However, in the study by Hendricks *et al.* normotensive, pre-hypertensive, and hypertensive patients treated for a long term with phentermine had no statistically significant increase in heart rate. Also, they "exhibited reductions in SBP and DBP that were similar to those in patients losing weight on a low-carbohydrate ketogenic diet alone" [169]. In addition, clinical trials using phentermine for 12 weeks have shown either reductions in heart rate and BP with phentermine or no changes in BP with phentermine administration [166-168]. A study of patients taking phentermine for weight loss (n=269) demonstrated "no abuse potential or amphetamine-like withdrawal upon cessation, even at doses higher than the commonly recommended and after treatment duration of up to 21 years" [170]. "Common side adverse effects for short-term (12-14 weeks) use include dry mouth, dizziness, fatigue, insomnia, headache, tachycardia, and palpitations" [171]. It should be mentioned that "there were reports of valvular heart disease when phentermine was used in combination with fenfluramine or dexfenfluramine" [172, 173]. "Of the 132 reports, 113 (86 %) met the cardiac valvulopathy case definition [173]. Of these 113 cases, 2 used fenfluramine alone, 16 (14 %) dexfenfluramine alone, 89 a combination of fenfluramine and phentermine, and 6 a combination of all three drugs. There were no cases reported with phentermine monotherapy use" [173].

PHENTERMINE/TOPIRAMATE COMBINATION

Based on the fact that weight loss has been accomplished with phentermine and topiramate *per se*, "a combination of phentermine-topiramate extended release was developed for once-daily oral dosing to enhance weight loss and to improve weight related comorbidities in overweight and/or obese patients" [174, 175]. In this low-dose formulation, phentermine is "readily absorbed and immediately released with early in the morning effect, whereas topiramate extended release is released later and provides effects through later periods of the day" [176]. "Phentermine-topiramate extended release, is administered as a once-daily capsule in 4 fixed-dose combinations: 3.75 mg phentermine/ 23 mg topiramate (starting dose); 7.5 mg phentermine/46 mg topiramate (recommended dose); 11.25 mg phentermine/69 mg topiramate; and 15 mg phentermine/92 mg topiramate (top dose). Dosage is increased over 14 days to 7.5 mg phentermine/46mg topiramate, with 25 additional titration to the top dose if weight loss is not adequate [177]. Weight loss should be re -evaluated after a period of 12 weeks, and if 5 % weight loss has not been achieved, therapy with phentermine-topiramate ER should be withdrawn" [178]. Recently, it has been demonstrated that phentermine-topiramate-ER achieved more weight loss than monotherapy with either agent and at lower doses of either agent, too [179]. "The efficacy, safety, and tolerability of phentermine-topiramate-ER were assessed in overweight and obese patients with co-morbidities in two one -year, randomized, placebo controlled phase 3 trials, the EQUIP and the CONQUER studies" [180-182]. "EQUIP included 1267 severely obese adults, 18-70 years of age, BMI $\geq$ 35 kg/m^2, without diabetes [183]. 40 % of participants withdrew from the study. At the top dose, mean one year weight loss was 10.9 % *versus* 1.6 % of initial weight for placebo. 67 % of patients, who were given the top dose, lost $\geq$ 5 % of their initial weight and 47 % lost $\geq$ 10 % of their initial weight, compared with 17 % and 7 % respectively in the placebo groups" [180].

"CONQUER included 2487 overweight and obese adults (18-70 years of age, BMI $\geq$ 27 kg/m^2 and $\leq$ 45 kg/m^2) with at least two weight- related comorbidities, including hypertension and type 2 diabetes mellitus [181]. 31 % of participants withdrew from the study. One year weight loss was 7.8 % with the recommended dose and 9.8 % with the top dose, *versus* 1.2 % with placebo. Besides, 62 % (recommended dose) and 70 % (top dose) lost $\geq$ 5 % of their initial weight *versus* 21 % for placebo, with 37 %, 48 %, and 7 % respectively losing $\geq$ 10 % of their initial weight" [181]. "SEQUEL, an extension to the CONQUER study, followed 78 % of CONQUER participants and who had completed the initial 56-week trial for a total of 108 weeks [182]. 84 % completed their second year of treatment with persistent weight loss of 9.3 % and 10.5 % at the recommended and top doses, respectively, *versus* 1.8 % for placebo, and revealed differences in many CVD risk factors. In addition, there was lower incidence of progression to type 2 diabetes mellitus in the top-dose group (0.9 %) *versus* the placebo group (3.7 %)"

[182]. "In a fourth study, adults with type 2 diabetes were evaluated in a 28-week extension of a 28-week placebo-controlled phase 2 trial (DM-230; 56 weeks total) [183]. A larger proportion of patients receiving phentermine-topiramate-ER also achieved at least 5 % weight loss after 56 weeks of treatment in the phase 2 and 3 studies, when compared with placebo" [184]. The most common adverse drug reactions include "paraesthesias, dizziness, dysgeusia, insomnia, constipation, and mouth dryness, while potential safety concerns include depression, anxiety, cognitive-related complaints (memory and attention), cardiovascular risk with a small increase of heart rate and teratogenicity" [184]. Given the fact that most users of obesity medications are women of reproductive age, the potential for teratogenicity is a great concern [184]. "A small increase in resting heart rate has been noticed in clinical trials of phentermine/topiramate-ER at higher doses, with more patients on top- dose (56.1 %) than placebo (42.1 %) having increases of more than 10 beats per minute, leading to concerns associated with its potential long- term effect on CVD events". The FDA approved the combination of phentermine/topiramate in 2012, based mainly on the strength of evidence from three clinical trials, "known as the CONQUER, EQUIP, and SEQUEL trials" [181-185]. Data from these three trials demonstrated that phentermine/topiramate was "efficacious in inducing and maintaining weight loss". From those three trials, approximately "75 % of treated subjects exhibited a 5 % weight loss, and approximately 50 % exhibited a 10 % weight loss". In addition, this weight loss was also accompanied by amelioration in serum lipid profile, waist circumference, and arterial blood pressure. "It is not clear if sustained weight loss requires indefinite treatment with phentermine/topiramate, but it has been suggested that discontinuation of the drug may lead to weight regain" [186]. Although phentermine/topiramate was associated with increased heart rates, the FDA decided that "the greater reductions in blood pressure and weight loss supported a favorable benefit-risk balance to gain FDA approval" [187]. Phentermine is an atypical amphetamine; as such, the more common actions of amphetamines such as CNS stimulation, elevation of blood pressure (BP), tachyphylaxis and tolerance were not noted in clinical trials with phentermine/topiramate ER at the studied doses [188]. The effect of phentermine/topiramate ER on the QTc interval was examined in a phentermine and topiramate-ER randomized, double-blind study, in which it was documented

that phentermine/topiramate-ER "did not affect cardiac repolarization as measured by the change from baseline in QTc" [189].

LORCASERIN

It was in 1948, when Rapport *et al.* found the chemical structure of serotonin to be 5-hydroxytryptamine [190]. Until then, serotonin was only known as a serum vasoconstrictive substance. Although the effects of serotonin on peripheral organs were already acknowledged, only when its' structural similarity to LSD was identified, serotonin and its receptors were related to neurotransmission [191]. The first distinction between serotonin receptor subtypes was proposed in 1957, with the "5- HT1 receptors" demonstrated to have "high affinity for serotonin" and the "5- HT2 receptors" to have "lower affinity for 5-HT". The 5-HT1 and 5-HT2 receptors were identified by Gaddum *et al.* in 1957 [192]. Nowadays, serotonin has been demonstrated to mediate its effects through "at least 14 different receptors" [193]. The serotonin 5-HT2 receptor sub- family comprises three distinct receptor subtypes, "5-HT2A, 5-HT2B, and 5-HT2C, which share remarkable homology and activate common signaling pathways, including Gqα- mediated stimulation of phospholipase-Cβ, elevation of intracellular inositol phosphates, and elevation of intracellular calcium" [194]. "The 5- HT2 family receptors couple to Gq G proteins and modulate phosphoinositide hydrolysis and PKC activation [195-198]. 5-HT2 family receptors have lower affinity for molecular serotonin than other 5-HT receptors and are often inhibited by both typical and atypical antipsychotic drugs" [199]. 5 -HT2A receptors represent the major serotonin receptors found in platelets, vascular smooth muscle, and the cerebral cortex. 5-HT2B receptors are mainly found in the heart, and drugs "likely mediate their actions on heart valves *via* a combination of G protein and β-arrestin signaling" [200]. For many decades now, it has been documented that 5-HT is an important neurotransmitter for the central regulation of appetite and that it is the 5-HT2C receptor, which mediates the anorectic actions of 5-HT. The anorectic actions of the withdrawn anti-obesity medication fenfluramine were due to the actions of its metabolite norfenfluramine at 5-HT2C receptors. The combination of "fenfluramine and the amphetamine derivative phentermine" was a highly effective appetite suppressant, but was withdrawn due to a very increased and thus, not acceptable rate of valvular heart disease. Fenfluramine was also related to pulmonary hypertension, although to a lesser extent than the incidence of valvular heart disease [201]. Nowadays, it is well known that "both valvular heart disease and pulmonary hypertension are likely due to off-target interactions of fenfluramine's principal metabolite norfenfluramine with the 5-HT2B serotonin receptor" [202-203]. Thus, "while fenfluramine is relatively inactive at 5-HT2B receptors, norfenfluramine potently activates 5-HT2B receptors" [204]. The main obstacle limiting the development of 5-HT2C agonists has been "the lack of selectivity over 5-HT2A and 5-HT2B receptors" [205]. Activation of 5-HT2A receptors may result in hallucinations, while the activation of 5-HT2B receptors has been involved in cardiac valvular disease and pulmonary hypertension associated with dexfenfluramine use [206-209]. Thus, discovery of selective 5-HT2C receptor agonists could lead to the development of a promising anti-obesity regimen, which has the potential to avoid adverse side effects, associated with older classes of non-selective serotoninergic modulators. Thus, Arena Pharmaceuticals focused on developing a "5-HT2C-selective agonist, with no 5-HT2B agonist activity at all", predicting that such a molecule would be a potential anorectic agent, without cardiovascular adverse side effects [210]. Lorcaserin was officially launched in the USA in June 2013 [211]. Lorcaserin or "[(1R)-(+)-8-chloro-2,3,4,5-tetrahydro-1-methyl-1H-3 benzazepine], is a selective and potent 5-HT2C agonist". It acts as a selective 5-HT2C agonist on pro-opiomelanocortin neurons, which results in releasing alpha melanocyte -stimulating-hormone (alpha-MSH). Alpha- MSH acts on MC4R in the paraventricular nucleus of the hypothalamus, causing a decrease in appetite [212, 213]. Lorcaserin satisfied these criteria and was encountered into clinical trials for combating obesity.

"Although the mechanisms of action of lorcaserin are not completely understood, it is believed to act as an agonist at central serotonin subtype 2C (5- HT2C) receptors located on hypothalamic pro- opiomelanocortin neurons" (Fig. **3**). Agonism of the 5-HT2C receptor leads to reduced food intake and increased satiety, in other words, to weight loss [214]. Lorcaserin is suggested to achieve its weight-reducing effects by decreasing caloric intake and increasing satiety and not by altering energy expenditure or substrate oxidation [215]. Lorcaserin is orally administered and can be taken irrespectively of meals. "Time to maximal serum concentrations is approximately 1.5-2 hours, and the elimination half-life is approximately 11 hours". Lorcaserin is "approximately 70 % protein-bound and undergoes hepatic metabolism to inactive metabolites, which are mainly eliminated *via* the kidneys". No dosing adjustments are required in patients with mild-to-moderate hepatic or renal impairment. However, clinical experience is lacking in patients with Child-Pugh class C hepatic impairment or a creatinine clearance between 30 mL/min and 50 mL/min. "It is not recommended to use lorcaserin in severe renal disease, ie creatinine clearance < 30 mL per minute" [216]. In 2010, the results from a large phase III clinical trial, the BLOOM trial were published. The BLOOM study "encountered 3182 subjects aged 18-65 years with a BMI of 30-45 kg/m^2 or with a BMI of 27-30 kg/m^2 plus one or more of the following: hypertension, cardiovascular disease, impaired glucose tolerance, dyslipidemia, or obstructive sleep apnea, showed that lorcaserin-treated individuals on 10 mg/d achieved, an average 5.8-kg weight loss compared with a 2.2-kg weight loss among placebo-treated patients (P < 0.001)" [217]. "47.5 % of 33 lorcaserin-treated patients achieved ≥ 5 % weight loss compared with 20.3 % for those on placebo. Furthermore, there was no evidence for an increase in clinically significant valvular heart disease in lorcaserin- treated patients compared with those receiving placebo" [218]. Similar results were reported from the BLOSSOM trial, which "enrolled 4008 individuals aged 18-65 years with a body mass index (BMI) of 30-45 kg/m^2 or a BMI of 27-29.9 kg/m^2 and at least one weight-related comorbidity, such as hypertension, cardiovascular disease, impaired glucose tolerance, dyslipidemia or obstructive sleep apnea. In this one year trial, two doses of lorcaserin were examined (10 or 20 mg/day); again, there was no difference in the incidence of valvular heart disease in placebo- and lorcaserin-treated individuals. After one year of treatment, at least 5 % weight loss was achieved by 47.2 %, 40.2 %, and 25 % in the lorcaserin twice daily, lorcaserin once daily, and placebo groups, respectively" [218]. "More than 10 % weight loss was achieved by 22.6 % of those receiving lorcaserin twice daily, 17.4 % of those receiving lorcaserin once daily, and 9.7 % of those receiving placebo [219]".

In addition, waist circumference decreased on lorcaserin treatment. Changes in low-density lipoprotein did not differ between the groups. HbA1c, systolic blood pressure, diastolic blood pressure and heart rate all showed minor and not significant decreases. In both trials, "headache, nausea, and dizziness were the most frequent adverse side effects, and there was no evidence for neuropsychiatric symptoms in either trial" [220]. Lorcaserin has also been evaluated in obese individuals with type 2 diabetes. In a pivotal study, the BLOOM-DM study, O'Neill *et al.* found that lorcaserin reduced weight and improved glycemic control in individuals with type 2 diabetes mellitus compared with that observed in placebo-treated individuals. It must be mentioned that both groups received metformin or sulfonylurea (or both), together with diet and exercise. Thus, it could not be concluded without any further studies that lorcaserin alone would be effective in inducing weight loss in obese type 2 diabetic patients. In this study, echocardiograms were performed, and no significant adverse side effect was noted with regards of the incidence of valvular heart disease [221]. These results suggest that lorcaserin in combination with diet and exercise can promote weight loss and, thus, might improve glycemic control in obese type 2 diabetic patients. However, lorcaserin alone, induces only modest weight loss, as pointed out in a recent meta-analysis, based on five studies, with a mean weight loss of 3.2 kg at 1 year. "Nasopharyngitis, sinusitis, and upper respiratory infections were more commonly observed among lorcaserin-treated individuals in one trial, but not in another". "Cancer risk in animal studies had been a concern of the FDA, when it rejected lorcaserin the first time, but they concluded that the incidence of mammary adenocarcinoma in female rats increased at plasma concentrations 87 times the daily human clinical dose". The incidence of "mammary fibroadenoma" was increased in female rats, but the significance for human beings remains unknown [222]. Other concern related to lorcaserin is "the potential for interaction with other selective serotonin reuptake inhibitors (*e.g.* many anti-depressants) or with monoamine oxidase inhibitors to cause serotonin syndrome and neuroleptic malignant syndrome".

Lorcaserin is contraindicated in pregnancy due to an increased risk of teratogenicity. "The future development of lorcaserin in Europe remains to be seen" [222].

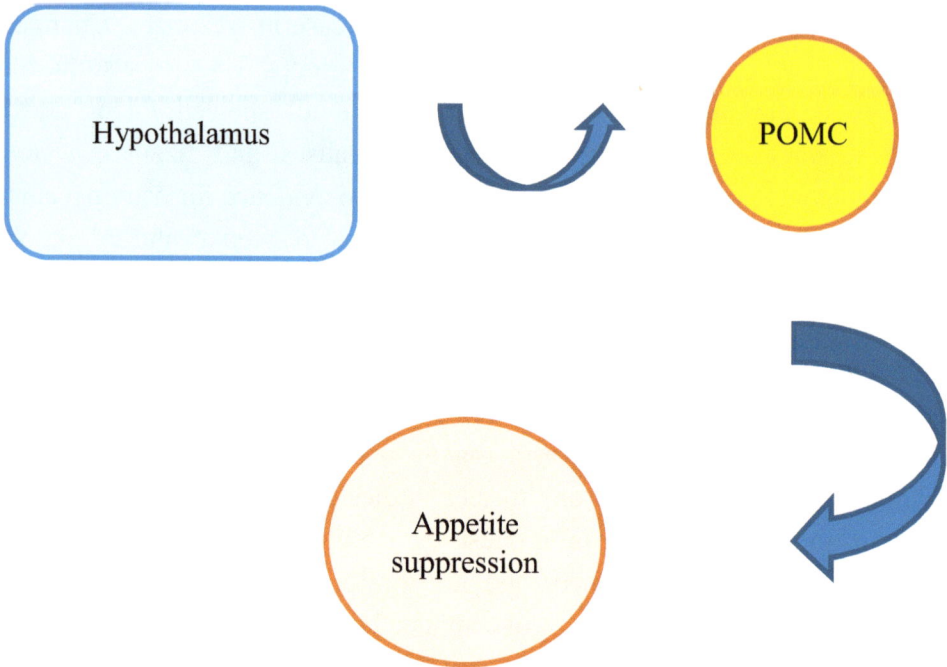

Figure 3: Lorcaserin exerts its effects by acting as an agonist of 5-HT2C receptors, which are located in the pro-opiomelanocortin neurons (POMC) of the Hypothalamus, resulting in appetite suppression.

OXYTOCIN

The story of oxytocin (OXT) began long ago, when it was recognized as a neuro-hypophyseal hormone, involved in lactation and uterine contraction. Nowadays, OXT has been shown to promote body weight lowering, by means of acting on "brain regulatory regions;" specifically, "the hypothalamus". Several studies have documented that OXT may play important roles in "affecting glucose metabolism as well as insulin secretion and lipolysis; functions being regulated both centrally and peripherally [223]". Few data have proposed that obesity is related to defects in OXT signaling. In particular, rodents with haplo- insufficiency of "single-minded 1 (SIM1) gene" have been shown to have low OXT expression and to be highly susceptible to diet-induced obesity [224]. In humans, mutations in SIM1 have been related to obesity; also, reduction in the number and size of OXT neurons have been found in patients with Prader-Willi syndrome, a genetic disorder characterized by obesity [224]. "The importance of OXT release dysregulation in feeding, body weight and obesity pathogenesis was demonstrated only recently, in a study using rodents model with high-fat diet (HFD)-induced obesity" [225]. In this study, HFD-induced obesity was mainly attributed to down-regulation of OXT release, through "increased vesicle binding of atypical synaptic exocytosis modulator, synaptotagmin-4 (Syt4)". Syt4, which was documented to be rich in OXT neurons of the hypothalamus, is "a member of vesicle fusion machinery; being Ca2+ insensitive, it is suggested to exert negative regulator properties on neurotransmitter or neuropeptide release" [226-227]. Zhang *et al.* recently reported that "HFD feeding led to up-regulation of Syt4 in hypothalamic OXT neurons, and Syt4 knockdown may restore OXT release to protect against HFD-induced obesity" [228]. While these results were indicative of the Syt4-mediated reduction of OXT release as a causal factor for the hypothalamic mechanism of obesity, the investigators also evaluated, whether OXT could be therapeutically useful in the treatment of obesity. Data were reported, documenting that administering OXT, provided protection against HFD-induced obesity in mice models, and that this protective ability was mainly caused by "a better regulation of feeding, energy expenditure and body weight". In addition, Deblon *et al.* reported that "chronic central OXT infusion, restricted weight gain dose- dependently and increased adipose tissue lipolysis in rats with HFD-induced obesity" [229]. Of note, Zhang *et al.* confirmed that peripheral administration of OXT may stimulate the central release of OXT, to account for the desired anti- obesity effects in rodents [230]. Furthermore, Morton *et al.* reported that "peripheral OXT delivery produced weight loss in rats with HFD-induced obesity, through reversal of leptin resistance by OXT's action on the satiety-related inputs in the brain" [231]. Moreover, recent therapeutic success was noted, when "OXT nasal spray effectively reduced weight and produced metabolic improvement in human patients with obesity, thus, further indicating OXT as a tempting drug target for combating obesity and its complications [232]". Apart from the native form, OXT-derived analogues have been found to effectively reduce body weight and glucose, too.

A very recent clinical trial using OXT to treat human obesity has yielded extremely encouraging results. In this clinical setting, researchers considered that "although OXT's effects on weight control are predicted to be largely mediated through the

CNS, peripheral OXT administration may stimulate the central release of OXT to induce metabolic benefits" [233]. Based on this notion, a clinical study was carried out to assess whether OXT delivery through nasal spray would be able to treat metabolic problems in patients with obesity. "OXT therapy in humans through intranasal route is already in practice for improving neuropsychiatric symptoms" [233, 234]. Besides, intranasal delivery route could make it more easily to cross the blood brain barrier and thus, stimulate the central neural circuitry. Zhang *et al.* found that multiple nasal sprays of OXT daily successfully decreased body weight in obese patients when compared to placebo [235]. The therapeutic benefit improved more by increasing duration treatment from 4 weeks to 8 weeks [235]. As it was also shown, "the effect of weight loss by OXT treatment was reflected by decreases in waist and hip circumferences of patients".

Altogether, all recent advances in studying OXT and its metabolic profile have provided us with promising results, for the development of OXT peptidyl drugs for treating obesity.

PPAR-γ

Similar to other nuclear receptors, PPAR-γ activity is regulated at "the post-translational level by modifications at key residues" [236, 237]. More specifically, obesity has been documented to "promote the phosphorylation of PPAR-γ at S273, and this post-translational modification is associated with the dysregulation of a subset of PPAR-γ target genes, many of which are dys-regulated in obesity". "pS273 sensitive genes include the insulin-sensitizing adipokines, such as adiponectin and adipsin [238]". "Partial agonist SPPARMγs blocks pS273 to the same extent as full agonist thiazolidenidiones". It has been demonstrated that there is a strong correlation in both obese rodents and obese humans regarding the anti-diabetic properties of drug administration and the decrease in pS273. These results suggest that efficacy may be due to the blockage of this "post-translational modulation and not from activation of the receptor itself, which partially explains the better side effect profile of the partial agonist SPPARγMs". Based on this finding, development of compounds, which were non-agonists of the receptor, but had the ability to block pS273 was further attained, such as SR1664. SR1664 has been documented to be efficacious as thiazolidenidiones in correcting increased serum glucose and fasting insulin levels in diabetic rats. In addition, SR1664 was anti-adipogenic and did not result in weight gain in the same rodents [239, 240]. While SR1664 offers hope for targeting PPAR-γ with improved therapeutic results, the pharmaceutical properties of this compound demand significant improvement. It is reasonable that "such neutral PPAR-γ antagonists, which have the ability to block pS273, will retain the transcriptional repressive properties of thiazolidenidiones in the macrophage, thus, maintaining the ability to reduce the inflammation that accompanies obesity". This compound would normalize expression of adipokines repressed by pS273, ameliorate insulin receptor signaling, without promoting adipogenesis and fluid retention, and would be neutral regarding bone density.

OREXIN

Recent data has reported the existence of a new class of neuropeptides, called orexins or hypocretins, which are produced in the hypothalamus, and whose actions are mediated by two types of receptors: 42 OX1R and OX2R. More specifically, the orexinergic neurons have been located exclusively in the lateral, dorsomedial and perifornical areas of the hypothalamus. Despite this highly specific anatomical origin, the orexinergic neurons are projected widely into a number of brainstem, cortical and limbic regions.

Orexin is a unique neuropeptide, which "promotes both feeding and energy expenditure, thus resulting in inducing weight loss". Several studies have investigated the relationship between circulating orexin and fat mass and have documented a "strong relationship between low plasma orexin and obesity" [241, 242]. Bronsky *et al.* showed that "plasma orexin increases with weight loss in obese children, suggesting that orexin may have a unique role in body weight balance: it provides a mechanism to burn calories through thermogenesis under conditions of caloric excess, arguably to increase fitness [243]". There is a need to develop strategies that "not only activate brown fat selectively, but also induce maturation in pre- adipocytes of obese subjects". Because brown pre-adipocytes express OXR1, orexin may not only be able to produce brown pre-adipocyte commitment, but also drive maturation of these precursors and resolve obesity by correcting BAT dysfunction, which is among the cornerstones of obesity [244, 245]. Other novel drugs, which have been explored and seem promising against obesity, in early clinical development, include beloranib, tesofensine and velerenib. Tesofensine acts by inhibiting noradrenaline, dopamine, and serotonin. The initial use of tesofensine in patients with Parkinson's or Alzheime's disease was observed to induce weight loss. Thus, tesofensine was developed for treating obesity. A 24-week phase II study with tesofensine demonstrated significant dose-dependent weight loss. The most common adverse events were dry mouth and nausea. However, "tesofensine 1.0 mg increased anger and hostility and both 0.5 mg and 1.0 mg of tesofensine increased the risk of confusion". Apart from that, serious adverse events did not differ significantly between tesofensine and placebo groups in the above-mentioned study [246, 247]. In theory, "drugs that increase dopamine levels have stimulant effects and may be associated with abuse". "However, the absorption and elimination of oral tesofensine in humans are slow". Besides, further data suggest that the abusive potential of tesofensine seems to be low in humans. Velneperit is an oral, small-molecule NPY5 receptor antagonist, which is administered once-daily. "Neuropeptide Y (NPY) is an orexigenic signaling molecule, which has a key-role in the regulation of energy balance, especially under conditions of food deprivation or reduced weight". Velneperit was designed to suppress elevated NPY levels, thus, promoting weight reduction [248]. Beloranib, an "analog of the natural substance fumagillin, is a methionine amino peptidase 2 inhibitor (MetAP2) [249]". "MetAP2's ability to suppress activity of extracellular regulated kinases 1 and 2 (ERK1/2) represents one of the key mechanisms for the anti- obesity properties [250]. MetAP2 inhibition results in suppression of sterol regulatory element binding protein (SREBP) activity, leading to reduced lipid and cholesterol biosynthesis *via* ERK-related pathways" [251, 252]. At the beginning, it was designed as an angiogenesis inhibitor for combating cancer. However, once the anti-obesity properties of MetAP2 inhibition became apparent, the clinical development began to focus on these properties and beloranib has been observed to possess promising results in preliminary clinical trials for treating obesity, too. The first Phase I study demonstrated that "beloranib (ZGN-433), at a dose of 0.9 mg/ m^2 was well tolerated and decreased body weight by a median of 1 kg per week and 3.1 % over 26 days, when compared to placebo, in severely obese subjects [253]". The results of the study also showed a "decline in hunger as well as changes in lipid parameters following treatment at 0.9 mg/m^2, such as a 38 % reduction in triglyceride levels and a 23 % reduction in LDL cholesterol levels". The second Phase I trial was a study to evaluate "the safety, tolerability, and efficacy of twice-weekly intravenously administered beloranib in severely obese women with a BMI of 39.1±3.7 kg/m². Individuals received 0.9 mg/m², 3 mg/m², or 6 mg/m² beloranib (n=17) or placebo (n=11) twice weekly over a 25-day period. The trial involved 34 subjects of whom 28 completed the study". Beloranib treatment for 25 days resulted in greater weight reduction *versus* the placebo-controlled group. Besides, subjects treated with beloranib had improved triglycerides, LDL cholesterol, waist circumference, diastolic blood pressure, and C-reactive protein, whereas there was no change in the placebo group [254]. The third Phase I trial was a randomized, double-blind study to evaluate the "safety, tolerability, and efficacy of intravenously administered ZGN-433 in severely obese women with a BMI of 37.8±0.6 kg/m² [254]. Individuals received ZGN-433 at 0.1 mg/m², 0.3 mg/m², or 0.9 mg/ m² or placebo twice weekly by intravenous administration over a 4-week period. The study encountered 31 subjects and 26 people completed it [255]. After 26 days of treatment, subjects had decreased LDL cholesterol levels by 22 % in the group that received 0.9 mg/m² of ZGN-433 *versus* a 2 % increase in the placebo group". On November 15, 2013, Zafgen Inc., announced body weight reduction from a Phase II study of beloranib for obesity. The 12-week study results demonstrated significant weight reduction and improvements in cardiometabolic risk markers in 147 obese individuals over 12 weeks of treatment, the largest and longest trial to date for the beloranib program [256]. In this study, Zafgen Inc., investigated the "safety, pharmacokinetics, and metabolic effects of beloranib in obese men and women. The trial evaluated 147 patients, of whom 122 completed the study. Subjects were mostly obese women with a mean body weight of 100.9 kg, and a mean BMI of 37.6 kg/m², who were enrolled into one of the four arms of the trial: n=37 for 0.6 mg; 37 for 1.2 mg; 35 for 2.4 mg; and n=38 in the placebo arm. Results showed that after 12 weeks of treatment, subjects on 0.6 mg, 1.2 mg, or 2.4 mg of beloranib lost an average of 5.5±0.5 kg, 6.9±0.6 kg, and 10.9±1.1 kg, respectively, *versus* losses of 0.4±0.4 kg for those in the placebo group". The results have also advocated that weight loss with beloranib was progressive and continuing at week 12. As previously observed, this study also revealed improved serum LDL-cholesterol levels, high density lipoprotein- cholesterol, triglycerides, and a reduction in blood pressure. The most common "adverse events with beloranib were nausea, diarrhea, headache, injection site bruising, and insomnia [257]".

Other anti-obesity agents

Brown adipose tissue (BAT) possesses too many vessels [258]. Thus, modulation of adipose tissue function by controlling angiogenesis may serve as a research tool for combating obesity. It is by means of these precursor cells in the vessel walls that adipogenesis is controlled by angiogenesis. Adipose- derived stem cells produce peptides, like "vascular endothelial growth factor (VEGF), fibroblast growth factor-2 (FGF-2), hepatocyte growth factor (HGF), placental growth factor (PlGF), secreted protein acidic and rich in cysteine (SPARC)/osteonectin, angiopoietins, matrix metalloproteinases (MMPs), leptin, resistin, and visfatin", which promote angiogenesis [259]. The VEGF-VEGFR-2 system regulates angiogenesis in BAT. Another mechanism, responsible for the stimulation of adipose angiogenesis, is the sympathetic release of neuropeptide Y (NPY). "NPY stimulates NPY- Y2 receptors, which activate adipose angiogenesis in BAT [260]". It has also been demonstrated that specific pharmacological inhibition of VEGFR-2 results in arresting adipose tissue expansion, mainly through inhibition of angiogenesis [261, 262]. Taken together, the above-mentioned studies suggest that inhibiting angiogenesis in white adipose tissue and BAT, may serve as a potential and fruitful anti-obesity regimen [263].

Novel targets, such as "histamine H3 receptor, VEGF, matrix-metalloproteinase and sirtuin receptors" - resveratrol playing a key-role in the latter- are also being investigated [264-265]. Other anti-obesity agents orlistat and cetilistat are gastrointestinal and pancreatic lipase inhibitors that reduce fat absorption. In a multicenter, randomized study "involving 869 subjects to determine the efficacy and safety of cetilistat (40, 80, or 120 mg TID) and orlistat (120 mg TID) in comparison to placebo, in obese patients with type 2 diabetes on metformin, for 12 weeks, similar reductions in body weight were observed in patients receiving cetilistat 80 or 120 mg TID or 120 mg TID orlistat; these reductions were significant *vs.* placebo (3.85 kg, 4.32 kg, and 3.78 kg, respectively) [266]". The results are comparable to those of orlistat, with approximately 30 % of the treated patients experiencing > 5 % weight loss compared to 19 % in the placebo group.

"The adverse side effects of cetilistat were similar to those reported with orlistat, although such events were less frequent, suggesting a better tolerability and therefore perhaps a better compliance [267]".

Bupropion stimulates "hypothalamic pro- opiomelanocortin (POMC) neurons that release alpha- melanocyte stimulating hormone (a-MSH) which binds to melanocortin-4 receptors, and thus, favors an anorexigenic action". "When a-MSH is released, POMC neurons release b- endorphin, an endogenous agonist of the mu-opioid receptor. Binding of b- endorphin to m-opioid receptors on POMC neurons mediates a negative feedback loop on POMC neurons leading to a decrease in the release of a- MSH. Blocking this inhibitory feedback loop with naltrexone is thought to facilitate a more potent and longer-lasting activation of POMC neurons, thereby amplifying its effects on energy balance. As a result, co- administration of bupropion and naltrexone produces a substantially greater effect on the POMC neurons, suggesting that the drugs act synergistically". Bupropion was combined with naltrexone in its sustained release form (Contrave™). Several phase III trials, "grouped under the Contrave Obesity Research (COR), have been conducted in both diabetic and non- diabetic patients: COR-I, COR-II, COR- BMOD and COR-Diabetes [268-269]. The naltrexone/bupropion patients lost significantly more weight (5.0 % *versus* 1.5 %, p < 0.001) at 56 weeks, with 45 % of patients achieving ≥ 5 % body weight loss, compared to 19 % with placebo. This combination resulted in improvements in depressive symptoms in addition to weight loss, as well as in a satisfactory recovery of eating-control in overweight and obese women with major depression". This combination has been well tolerated among most patients; nausea was the most frequently reported adverse event, which was associated to higher naltrexone doses. The FDA decided the approval of contrave (naltrexone hydrochloride and bupropion hydrochloride extended-release tablets) as treatment option for chronic weight management. The drug was approved for use in adults with a body mass index (BMI) of 30 or greater (obesity) or adults with a BMI of 27 or greater (overweight), who have at least one weight-related condition, such as high blood pressure, type 2 diabetes mellitus, or dyslipidemia.

The combination of bupropion with the antiepileptic agent zonisamide has been evaluated in phase II trials. Zonisamide's mechanism of action has not been fully understood; however, it seems to possess a "biphasic dopamine and serotoninergic activity". "A 24-week randomized, control trial of bupropion 300 mg combined with zonisamide 400 mg, achieved a greater weight loss (9.2 %) than either drug alone (bupropion 6.6 %, zonisamide 3.6 %) or placebo (0.4 %). Weight loss with zonisamide and bupropion appeared to be greater than that observed with the bupropion/naltrexone combination over the same period of treatment [270, 271]".

CONCLUSION

Due to the obesity epidemic, there is ongoing research regarding novel classes of anti-obesity drugs, such as GLP-1 receptor agonists, 5- HT2C selective serotonin receptor agonists, oxyntomodulin, oxytocin, orexin and PPARγ's (Fig. **4**). These new classes of anti-obesity agents seem to be very promising in our battle against the obesity tide.

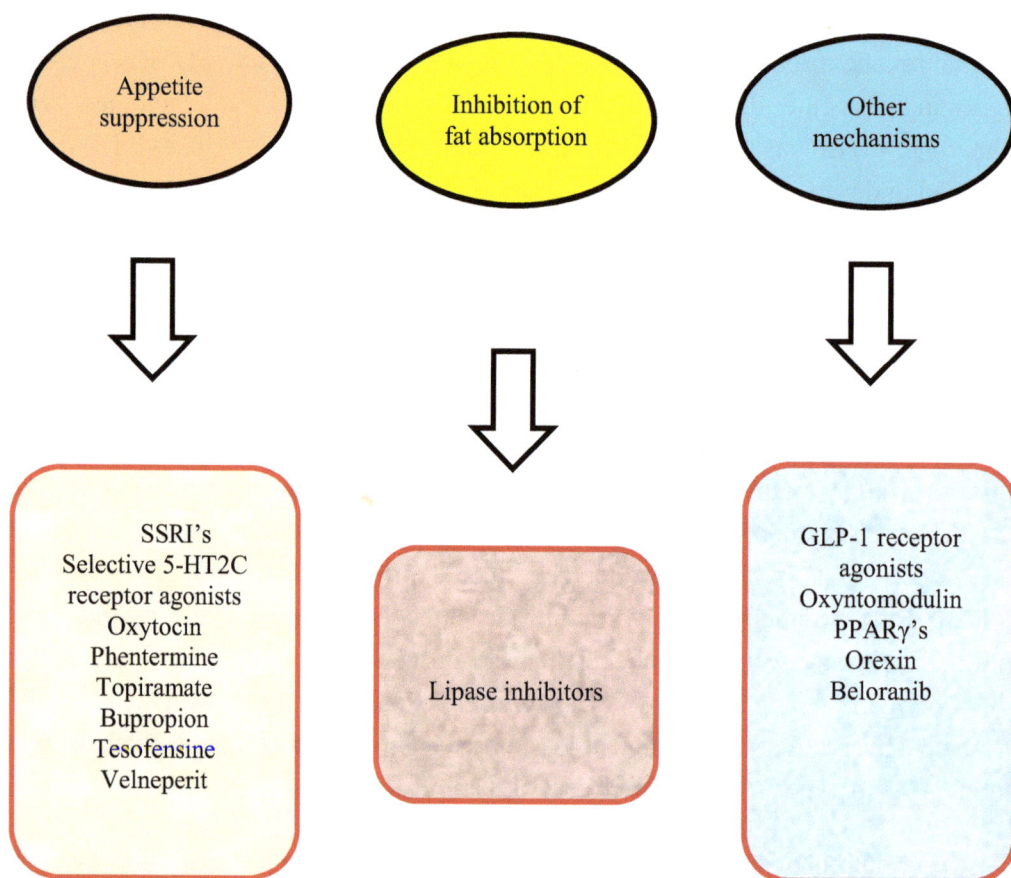

Figure 4: Mechanisms of actions of anti-obesity drugs under discovery.

Table 1: Anti-obesity drugs: the past, the present and the future.

Past	Present	Future
Thyroid preparations	Orlistat	Cetilista
Amphetamine	Phentermine	Tesofensine
Phentermine	Lorcaserin	Empatic (bupropion+zonisamide)
Sibutramine	Qsymia (phentermine+topiramate ER)	Beloranib
Rimonabant	Contrave (naltrexone+bupropion ER)	Velneperit
		Resveratrol
		VEGF/VEFGR2
		MMPs

AKNOWLEDGEMENTS

Declared None.

CONFLICT OF INTEREST

The authors confirm that this article content has no conflict of interest.

ABBREVIATIONS

5-HT1	=	5 Hydroxytryptamine 1
5-HT2C Receptor	=	5-Hydroxytryptamine 2C Receptor
α-MSH	=	Alpha melanocyte-stimulating-hormone
BAT	=	Brown adipose tissue
BMI	=	Body mass index
BP	=	Blood pressure
CNS	=	Central nervous system
CT	=	Computer tomography

CVD	=	Cardiovascular disease
DBP	–	Diastolic blood pressure
DPP-4	=	Dipeptyl-peptidase-4
ER	=	Extended release
ERK1/2	=	Extracellular signal regulated kinases 1 and 2
FDA	=	Food and drug administration
GABA	=	Gamma aminobutyric acid
GCG	=	Glucagon
GCGR	=	Glucagon receptor
GLP1-1	=	Glucagon like peptide-1
GLP-1R	=	Glucagon like peptide-1 receptor
HCL	=	Hydrochloride
HFD	=	High fat diet
IP-1	=	Intervening peptide-1
LSD	=	Lysergic acid diethylamide
MetAP2	=	Methionine amino peptidase 2 inhibitor
NAc	=	Nucleus accumbens
NPY	=	Neuropeptide Y
OXM	=	Oxyntomodulin
OXR1	=	Orexin receptor 1

OXR2   = Orexin receptor 2

OXT   = Oxytocin

PC-2   = Pro-hormone convertase-2

PKC   = Protein kinase C

POMC   = Pro-opiomelanocortin

PPaR-γ = Peroxisome proliferator agonist receptor γ

PYY   = Peptide YY

RYGB   = Roux en Y gastric bypass

SBP   = Systolic blood pressure

SIM1   = single-minded 1

SP-1   = Spacer peptide-1

Syt-4   = Synaptotagmin-4

T2D   = Type 2 diabetes

VEGF   = Vascular endothelial growth factor

VEGFR   = Vacular endothelial growth factor receptor

VTA   = Ventral tegmental areas

REFERENCES

[1]   Heaton JM. The distribution of brown adipose tissue in the human. J Anat 1972; 112: 35-9.
[2]   Merklin RJ. Growth and distribution of human fetal brown fat. Anat Rec 1974; 178: 637-45.
[3]   Cunningham S, Leslie P, Hopwood D, *et al.* The characterization and energetic potential of brown adipose tissue in man. Clin Sci (Lond) 1985; 69: 343-8.
[4]   Lean ME. Brown adipose tissue in humans. Proc Nutr Soc 1989; 48: 243-56.

[5]   Nedergaard J, Bengtsson T, Cannon B. Unexpected evidence for active brown adipose tissue in adult humans. Am J Physiol Endocrinol Metab 2007; 293: E444-452.

[6]   Cypess AM, Lehman S, Williams G, et al. Identification and importance of brown adipose tissue in adult humans. N Engl J Med 2009; 360: 1509-1517.

[7]   van Marken Lichtenbelt WD, Vanhommerig JW, Smulders NM, et al. Cold-activated brown adipose tissue in healthy men. N Engl J Med 2009; 360: 1500-1508.

[8]   Virtanen KA, Lidell ME, Orava J, et al. Functional brown adipose tissue in healthy adults. N Engl J Med 2009; 360: 1518-1525.

[9]   Betz MJ, Slawik M, Lidell ME, et al. Presence of brown adipocytes in retroperitoneal fat from patients with benign adrenal tumors: relationship with outdoor temperature. J Clin Endocrinol Metab 2013; 98: 4097-4104.

[10]   Lee P, Zhao JT, Swarbrick MM, et al. High prevalence of brown adipose tissue in adult humans. J Clin Endocrinol Metab 2011; 96: 2450-2455.

[11]   Saito M, Okamatsu-Ogura Y, Matsushita M, et al. High incidence of metabolically active brown adipose tissue in healthy adult humans: effects of cold exposure and adiposity. Diabetes 2009; 58: 1526-1531.

[12]   Lee P, Greenfield JR, Ho KK, Fulham MJ. A critical appraisal of the prevalence and metabolic significance of brown adipose tissue in adult humans. Am J Physiol Endocrinol Metab 2010; 299: E601-606.

[13]   Rothwell NJ, Stock MJ, Stribling D. Diet-induced thermogenesis. Pharmacol Ther 1982; 17: 251-268.

[14]   Stock MJ. Thermogenesis and brown fat: relevance to human obesity. Infusionstherapie 1989; 16: 282-284.

[15]   Rothwell NJ, Stock MJ. Influence of noradrenaline on blood flow to brown adipose tissue in rats exhibiting diet-induced thermogenesis. Pflugers Arch 1981; 389: 237-242.

[16]   Feldmann HM, Golozoubova V, Cannon B, Nedergaard J. UCP1 ablation induces obesity and abolishes diet-induced thermogenesis in mice exempt from thermal stress by living at thermoneutrality. Cell Metab 2009; 9: 203-209.

[17]   Verty ANA, Allen AM, Oldfield BJ. The effects of rimonabant on brown adipose tissue in rat: implications for energy expenditure. Obesity (Silver Spring) 2009; 17: 254-261.

[18]   Bartness TJ, Vaughan CH, Song CK. Sympathetic and sensory innervation of brown adipose tissue. Int J Obes (Lond) 2010; 34(Suppl 1): S36-42.

[19]   Whittle AJ, López M, Vidal-Puig A. Using brown adipose tissue to treat obesity - the central issue. Trends Mol Med 2011; 17: 405-411.

[20]   Morrison SF, Madden CJ, Tupone D. Central control of brown adipose tissue thermogenesis. Front Endocrinol (Lausanne) 2012; 3; PMID: 22389645.

[21]   Lockie SH, Heppner KM, Chaudhary N, et al. Direct control of brown adipose tissue thermogenesis by central nervous system glucagon-like peptide-1 receptor signaling. Diabetes 2012; 61: 2753-2762.

[22]   Ghatei MA, Uttenthal LO, Christofides ND, Bryant MG, Bloom SR. Molecular forms of human enteroglucagon in tissue and plasma: plasma responses to nutrient stimuli in health and in disorders of the upper gastrointestinal tract. J Clin Endocrinol Metab 1983; 57: 488-495.

[23]   Le Quellec A, Clapie M, Callamand P, et al. Circulating oxyntomodulin-like immunoreactivity in healthy children and children with celiac disease. J Pediatr Gastroenterol Nutr 1998; 27: 513-518.

[24]   Holst JJ. Enteroglucagon. Ann Rev Physiol 1998; 59: 257-271.

[25]   Drucker DJ. Biologic actions and therapeutic potential of the proglucagon-derived peptides. Nature Clinical Practice. Endocrinol Metab 2005; 1: 22-31.

[26]   Campos RV, Lee YC, Drucker DJ. Divergent tissue-specific and developmental expression of receptors for glucagon and glucagon-like peptide-1 in the mouse. Endocrinology 1994; 134: 2156-2164.

[27]   Larsen PJ, Tang-Christensen M, Holst JJ, Orskov C. Distribution of glucagon-like peptide-1 and other preproglucagon-derived peptides in the rat hypothalamus and brainstem. Neuroscience 1997; 77: 257-270.

[28]   Rouille Y, Westermark G, Martin SK, Steiner DF. Proglucagon is processed to glucagon by prohormone convertase PC2 in a TC1-6 cells. PNAS 1994; 91: 3242-3246.

[29]   Kieffer TJ, Habener JF. The glucagon-like peptides. Endocrine Rev 1999; 20: 876-913.

[30]   Furuta M, Zhou A, Webb G, *et al*. Severe defect in proglucagon processing in islet A-cells of prohormone convertase 2 null mice. J Biol Chem 2001; 276: 27197-27202.

[31]   Brubaker PL. A beautiful cell (or two or three?) Endocrinology 2012; 153: 2945-2948.

[32]   Habib AM, Richards P, Cairns LS, *et al*. Overlap of endocrine hormone expression in the mouse intestine revealed by transcriptional profiling and flow cytometry. Endocrinology 2012; 153: 3054-3065.

[33]   Unger RH, Ohneda A, Valverde I, Eisentraut AM, Exton J. Characterization of the responses of circulating glucagon-like immunoreactivity to intraduodenal and intravenous administration of glucose. J Clin Invest 1968; 47: 48-65.

[34]   Valverde I, Rigopoulou D, Exton J, Ohneda A, Eisentraut A, Unger RH. Demonstration and characterization of a second fraction of glucagonlike immunoreactivity in jejunal extracts. Am J Med Sci 1968; 255: 415-420.

[35]   Bataille D, Gespach C, Coudray AM, Rosselin G. 'Enteroglucagon': a specific effect on gastric glands isolated from the rat fundus. Evidence for an 'oxyntomodulin' action. Bioscience Reports 1981; 1: 151-155.

[36]   Bataille D, Gespach C, Tatemoto K, *et al*. Bioactive enteroglucagon (oxyntomodulin): present knowledge on its chemical structure and its biological activities. Peptides 1981; 2: (Suppl 2): 41-44.

[37]   Bataille D, Tatemoto K, Coudray AM, Rosselin G, Mutt V. Bioactive 'enteroglucagon' (oxyntomodulin): evidence for a C-terminal extension of the glucagon molecule. Comptes Rendus des Seances de l'Academie des Sciences 1981; Serie III, Sciences de la Vie. 293: 323-328.

[38]   Dubrasquet M, Bataille D, Gespach C. Oxyntomodulin (glucagon-37 or bioactive enteroglucagon): a potent inhibitor of pentagastrin-stimulated acid secretion in rats. Biosci Rep 1982; 2: 391-395.

[39]   Thim L, Moody AJ. The amino acid sequence of porcine glicentin. Peptides 1981; 2 (Suppl 2): 37-39.

[40]   Sundby F, Jacobsen H, Moody AJ. Purification and characterization of a protein from porcine gut with glucagon-like immunoreactivity. Hormone Metab Res 1976; 8: 366-371.

[41]   Holst JJ. Evidence that enteroglucagon (II) is identical with the C-terminal sequence (residues 33-69) of glicentin. Biochem J 1982; 207: 381-388.

[42]   Larsson LI, Holst JJ, Hakanson R, Sundler F. Distribution and properties of glucagon immunoreactivity in the digestive tract of various mammals: an immunohistochemical and immunochemical study. Histochemistry 1975; 44: 281-290.

[43]   Baldissera FG, Holst JJ. Glucagon-related peptides in the human gastrointestinal mucosa. Diabetologia 1984; 26: 223-228.

[44]   Koole C, Wootten D, Simms J, *et al.* Allosteric ligands of the glucagon-like peptide 1 receptor (GLP-1R) differentially modulate endogenous and exogenous peptide responses in a pathway-selective manner: implications for drug screening. Mol Pharmacol 2010; 78: 456-465.

[45]   Dakin CL, Small CJ, Batterham RL, *et al.* Peripheral oxyntomodulin reduces food intake and body weight gain in rats. Endocrinology 2004; 145: 2687-2695.

[46]   Wynne K, Park AJ, Small CJ, *et al.* Oxyntomodulin increases energy expenditure in addition to decreasing energy intake in overweight and obese humans: a randomised controlled trial. Intern J Obes 2006; 30: 1729-1736.

[47]   Turton MD, O'Shea D, Gunn I, *et al.* A role for glucagon-like peptide-1 in the central regulation of feeding. Nature 1996; 379: 69-72.

[48]   Sowden GL, Drucker DJ, Weinshenker D, Swoap SJ. Oxyntomodulin increases intrinsic heart rate in mice independent of the glucagon-like peptide-1 receptor. Am J Physiol 2007; 292: R962-R970.

[49]   Wynne K, Field BC, Bloom SR. The mechanism of action for oxyntomodulin in the regulation of obesity. Curr Opinion Investig Drugs 2010; 11: 1151-1157.

[50]   Schulman JL, Carleton JL, Whitney G, Whitehorn JC. Effect of glucagon on food intake and body weight in man. J Applied Physiol 1957; 11: 419-421.

[51]   Salter JM. Metabolic effects of glucagon in the Wistar rat. Am J Clin Nutr 1960; 8: 535-539.

[52]   Salter JM, Ezrin C, Laidlaw JC, Gornall AG. Metabolic effects of glucagon in human subjects. Metabolism 1960; 9: 753-768.

[53]   Penick SB, Hinkle LE Jr. Depression of food intake induced in healthy subjects by glucagon. N Engl J Med 1961; 264: 893-897.

[54]   Habegger KM, Heppner KM, Geary N, Bartness TJ, DiMarchi R, Tschop MH. The metabolic actions of glucagon revisited. Nature Rev 2010; Endocrinology. 6: 689-697.

[55]   Jones BJ, Tan T, Bloom SR. Minireview: glucagon in stress and energy homeostasis. Endocrinology 2012; 153: 1049-1054.

[56]   Engel SS, Xu L, Andryuk PJ, *et al.* Efficacy and tolerability of MK-0893, a glucagon receptor antagonist (GRA), in patients with type 2 diabetes (T2DM). Diabetes 2011; 60 (Suppl 1): A85.

[57]   Maida A, Lovshin JA, Baggio LL, Drucker DJ. The glucagon-like peptide-1 receptor agonist oxyntomodulin enhances b-cell function but does not inhibit gastric emptying in mice. Endocrinology 2008; 149: 5670-5678.

[58]   Parlevliet ET, Heijboer AC, Schroder-van der Elst JP, *et al.* Oxyntomodulin ameliorates glucose intolerance in mice fed a high-fat diet. Am J Physiol 2008; Endocrinology and Metabolism. 294 E142-E147.

[59]   Schjoldager BT, Baldissera FG, Mortensen PE, Holst JJ, Christiansen J. Oxyntomodulin: a potential hormone from the distal gut. Pharmacokinetics and effects on gastric acid and insulin secretion in man. Eur J Clin Invest 1988; 18: 499-503.

[60]   Day JW, OttawayN, Patterson JT, *et al.* A new glucagon and GLP-1 co-agonist eliminates obesity in rodents. Nature Chem Biol 2009; 5: 749-757.

[61]   Kerr BD, Flatt PR, Gault VA. (D-Ser2) Oxm (mPEG-PAL): a novel chemically modified analogue of oxyntomodulin with antihyperglycaemic, insulinotropic and anorexigenic actions. Biochemical Pharmacol 2010; 80: 1727-1735.

[62]   Liu YL, Ford HE, Druce MR, *et al.* Subcutaneous oxyntomodulin analogue administration reduces body weight in lean and obese rodents. Int J Obes 2010; 34: 1715-1725.

[63]   Santoprete A, Capito E, Carrington PE, *et al*. DPP-IV-resistant, long-acting oxyntomodulin derivatives. J Peptide Sci 2011; 17: 270-280.

[64]   Fosgerau KS, Larsen M, BaeK SA, Meier E, Groendahl C, Bak HH. Combination of long-acting insulin with the dual GluGLP-1 agonist ZP2929 causes improved glycemic control without body weight gain in db/db mice. Diabetes 2011; 60 (Suppl 1): A418. (doi: 10.2337/db11-1487-1624).

[65]   Han VK, Hynes MA, Jin C, Towle AC, Lauder JM, Lund PK. Cellular localization of proglucagon/glucagon-like peptide I messenger RNAs in rat brain. J Neurosci Res 1986; 16: 97-107.

[66]   Jin SL, Han VK, Simmons JG, Towle AC, Lauder JM, Lund PK. Distribution of glucagon-like peptide-I (GLP-I), glucagon, and glicentin in the rat brain: an immune-cytochemical study. J Comp Neurol 1988; 271: 519-532.

[67]   Larsen PJ, Tang-Christensen M, Holst JJ, Orskov C. Distribution of glucagon-likepeptide-1 and other preproglucagon-derived peptides in the rat hypothalamus and brainstem. Neuroscience 1997; 77: 257-270.

[68]   Reimann F, Habib AM, Tolhurst G, Parker HE, Rogers GJ, Gribble FM. Glucose sensing in L cells: a primary cell study. Cell Metab 2008; 8: 532-539.

[69]   Hayes MR, DeJonghe BC, Kanoski SE. Role of the glucagon-like-peptide-1receptor in the control of energy balance. Physiol Behav 2010; 100: 503-510.

[70]   Graham DL, Erreger K, Galli A, Stanwood GD. GLP-1 analog attenuates cocaine reward. Mol Psychiatry 2012; 106: 574-578.

[71]   Parkes DG, Mace KF, Trautmann ME. Discovery and development of exenatide: the first antidiabetic agent to leverage the multiple benefits of the incretin hormone, GLP-1. Expert Opin Drug Discov 2013; 8: 219-244.

[72]   Wang Y, Perfetti R, Greig NH, *et al*. Glucagon-like-peptide-1 can reverse the age-related decline in glucose tolerance in rats. J Clin Invest 1997; 99: 2883-2889.

[73]   Greig N H, Holloway HW, De Ore KA, *et al*. Once daily injection of exendin-4 to diabetic mice achieves longterm beneficial effects on blood glucose concentrations. Diabetologia 1999; 42: 45-50.

[74]   Agerso H, Jensen LB, Elbrond B, Rolan P, Zdravkovic M. The pharmacokinetics, pharmacodynamics, safety and tolerability of NN2211, a new long-acting GLP-1 derivative, in healthy men. Diabetologia 2002; 45: 195-202.

[75]   Drucker DJ, Dritselis A, Kirkpatrick P. Liraglutide. Nat Rev Drug Discov 2010; 9: 267-268.

[76]   Turton MD, O'Shea D, Gunn I, *et al*. A role for glucagon-like-peptide-1 in the central regulation of feeding. Nature 1996; 379: 69-72.

[77]   Naslund E, Barkeling B, King N, *et al*. Energy intake and appetite are suppressed by glucagon-like-peptide-1 (GLP-1) in obese men. Int J Obes Relat Metab Disord 1999; 23: 304-311.

[78]   Hayes MR, Skibicka KP, Grill HJ. Caudal brainstem processing issufficient for behavioral, sympathetic, and parasympathetic responses driven by peripheral and hindbrain glucagon-like-peptide-1 receptor stimulation. Endocrinology 2008; 149: 4059-4068.

[79]   Larsen PJ, Tang-Christensen M, Jessop DS. Central administration of glucagon-like-peptide-1 activates hypothalamic neuroendocrine neurons in the rat. Endocrinology 1997; 138: 4445-4455.

[80]   Astrup A, Carraro R, Finer N, *et al*. Safety, tolerability and sustained weight loss over 2 years with the once-daily human GLP-1 analog, liraglutide. Int J Obes (Lond.) 2012; 36: 843-854.

[81]   Astrup A, Rossner S, VanGaal L, *et al.* Effects of liraglutide in the treatment of obesity: a randomised, double-blind, placebo-controlled study. Lancet 2009; 374: 1606-1616.

[82]   Sandoval DA, Bagnol D, Woods SC, D'Alessio DA, Seeley RJ. Arcuate glucagon-like-peptide1 receptors regulate glucose homeostasis but not food intake. Diabetes 2008; 57: 2046-2054.

[83]   Trapp S, Hisadome K. Glucagon-like-peptide1 and the brain: central actions, central sources. Auton Neurosci 2011; 161: 14-19.

[84]   Dickson SL, Shirazi RH, Hansson C, Bergquist F, Nissbrandt H, Skibicka KP. The glucagon-like-peptide1 (GLP-1) analogue, exendin-4, decreases the rewarding value of food: a new role for mesolimbic GLP-1 receptors. J Neurosci 2012; 32: 4812-4820.

[85]   Wise RA. Dopamine and food reward: back to the elements. Am J Physiol Regul Integr Comp Physiol 2004; 286: R13.

[86]   Wise RA. Dopamine, learning and motivation. Nat Rev Neurosci 2004; 5: 483-494.

[87]   Wise RA. Role of brain dopamine in food reward and reinforcement. Philos Trans R Soc Lond Biol Sci 2006; 361: 1149-1158.

[88]   Fulton S, Pissios P, Manchon R, *et al.* Leptin regulation of the mesoaccumbens dopamine pathway. Neuron 2006; 51: 811-822.

[89]   Hommel JD, Trinko R, Sears RM, *et al.* Leptin receptor signalling in midbrain dopamine neurons regulates feeding. Neuron 2006; 51: 801-810.

[90]   Abizaid A. Ghrelin and dopamine: new insights on the peripheral regulation of appetite. J Neuroendocrinol 2009; 21: 787-793.

[91]   Abizaid A, Liu ZW, Andrews ZB, *et al.* Ghrelin modulates the activity and synaptic input organization of midbrain dopamine neurons while promoting appetite. J Clin Invest 2006; 116: 3220-3239.

[92]   DiLeone RJ, Taylor JR, Picciotto MR. The drive to eat: comparisons and distinctions between mechanisms of food reward and drug addiction. Nat Neurosci 2012; 15: 1330-1335.

[93]   Rinaman L. Ascending projections from the caudal visceral nucleus of the solitary tract to brain regions involved in food intake and energy expenditure. Brain Res 2010; 1350: 18-34.

[94]   Alhadeff AL, Rupprecht LE, Hayes MR. GLP-1 neurons in the nucleus of the solitary tract project directly to the ventral tegmental area and nucleus accumbens to control for food intake. Endocrinology 2012;153: 647-658.

[95]   Dossat AM, Diaz R, Gallo L, Panagos A, Kay K, Williams DL. Nucleus accumbens GLP-1 receptors influence meal size and palatability. Am J Physiol Endocrinol Metab 2013; 304: E1314-E1320.

[96]   Dossat AM, Lilly N, Kay K, Williams DL. Glucagon- like-peptide1 receptors in nucleus accumbens affect food intake. J Neurosci 2011; 31: 14453-14457.

[97]   Ussher JR, Drucker DJ. Cardiovascular biology of the incretin system. Endocr Rev 2012; 33(2): 187-215.

[98]   Baggio LL, Drucker DJ. Biology of incretins: GLP-1 and GIP. Gastroenterology 2007; 132: 2131-2157.

[99]   Drucker DJ, Nauck MA. The incretin system: glucagon-like peptide-1 receptor agonists and dipeptidyl peptidase-4 inhibitors in type 2 diabetes. Lancet 2006; 368: 1696-1705.

[100]   Nauck M, Stöckmann F, Ebert R, Creutzfeldt W. Reduced incretin effect in type 2 (non-insulin-dependent) diabetes. Diabetologia 1986; 29: 46-52.

[101]   Højberg PV, Vilsbøll T, Rabøl R, *et al.* Four weeks of near-normalisation of blood glucose improves the insulin response to glucagon-like peptide-1 and glucose dependent insulin-tropic polypeptide in patients with type 2 diabetes. Diabetologia 2009; 52: 199-207.

[102] Nauck MA, Kleine N, Orskov C, Holst JJ, Willms B, Creutzfeldt W. Normalization of fasting hyperglycaemia by exogenous glucagon-like peptide 1 (7-36 amide) in type 2 (non-insulin-dependent) diabetic patients. Diabetologia 1993; 36: 741-744.

[103] Neff LM, Kushner RF. Emerging role of GLP-1 receptor agonists in the treatment of obesity. Diab Metabol Syndr Obes 2010; 3: 263-273.

[104] Nauck MA, Ratner RE, Kapitza C, Berria R, Boldrin M, Balena R. Treatment with the human once-weekly glucagon-like-peptide-1 analogue taspoglutide in combination with metformin improves glycemic control and lowers body weight in patients with type 2 diabetes inadequately controlled with metformin alone. Diabetes Care 2009; 32: 1237-1243.

[105] Rosenstock J, Reusche J, Bush M, Yang F, Stewart M. Potential of albiglutide, a long acting GLP-1 receptor agonist in type 2 diabetes. Diabetes Care 2009; 32: 1880-1886.

[106] Drucker DJ, Buse JB, Taylor K, *et al*. Exenatide once weekly versus twice daily for the treatment of type 2 diabetes: a randomised, open-label, non-inferiority study. Lancet 2008; 372: 1240-1250.

[107] Russell-Jones D. Molecular, pharmacological and clinical aspects of liraglutide, a once-daily human GLP-1 analogue. Mol Cell Endocrinol 2009; 297: 137-140.

[108] Steensgaard DB, Thomsen JK, Olsen HB, Knudsen LB. The molecular basis for the delayed absorption of the once-daily human GLP-1 analogue, liraglutide. Diabetes 2008; 57(Suppl. 1): A164.

[109] Pratley RE, Nauck M, Bailey T, *et al*. Liraglutide versus sitagliptin for patients with type 2 diabetes who did not have adequate glycaemic control with metformin: a 26-week, randomised, parallel group, open-label trial. Lancet 2010; 375: 1447-1456.

[110] Umpierrez G, Blevins T, Rosenstock J, Cheng C, Bastyr E, Anderson J. The effect of LY2189265 (GLP-1 analogue) once weekly on HbA1c and beta cell function in uncontrolled type 2 diabetes mellitus: the EGO study analysis. Diabetologia 2009; 52(Suppl. 1): S59-S42.

[111] Pratley R, Nauck M, Bailey T, *et al*. One year of liraglutide treatment offers sustained and more effective glycaemic control and weight reduction compared with sitagliptin, both in combination with metformin, in patients with type 2 diabetes: a randomised, parallel-group, open-label trial. Int J Clin Pract 2011; 65: 397-407.

[112] Tatarkiewicz K, Smith PA, Sablan EJ, *et al*. Exenatide does not evoke pancreatitis and attenuates chemically induced pancreatitis in normal and diabetic rodents. Am J Physiol Endocrinol Metab 2010; 299: E1076-E1086.

[113] Engel SS, Williams-Herman DE, Golm GT, *et al*. Sitagliptin: review of preclinical and clinical data regarding incidence of pancreatitis. Int J Clin Pract 2010, 64: 984-990.

[114] Drucker DJ, Sherman SI, Bergenstal RM, Buse JB: The safety of incretin based therapies-review of the scientific evidence. J Clin Endocrinol Metab 2011; 96: 2027-2031.

[115] Dore DD, Bloomgren GL, Wenten M, *et al*. A cohort study of acute pancreatitis in relation to exenatide use. Diabetes Obes Metab 2011; 13: 559-566.

[116] Garg R, Chen W, Pendergrass M: Acute pancreatitis in type 2 diabetes treated with exenatide or sitagliptin: a retrospective observational pharmacy claims analysis. Diabetes Care 2010; 33: 2349-2354.

[117] Bjerre KL, Madsen LW, Andersen S, *et al*.: Glucagon-like peptide-1 receptor agonists activate rodent thyroid C-cells causing calcitonin release and C-cell proliferation. Endocrinology 2010; 151: 1473-1486.

[118] Parks M, Rosebraugh C: Weighing risks and benefits of liraglutide-the FDA's review of a new antidiabetic therapy. N Engl J Med 2010; 362: 774-777.

[119] Nisal K, Kela R, Khunti K, Davies M. Comparison of efficacy between incretin-based therapies for type 2 diabetes mellitus. BMC Medicine 2012; 10: 152.

[120] Owens DR. New horizons - alternative routes for insulin therapy. Nature Rev 2002; Drug Discovery. 1: 529-540.

[121] Sloop KW, Willard FS, Brenner MB, *et al*. Novel small molecule glucagon-like peptide-1 receptor agonist stimulates insulin secretion in rodents and from human islets. Diabetes 2010; 59: 3099-3107.

[122] Lee KC, Chae SY, Kim TH, Lee S, Lee ES, Youn YS. Intrapulmonary potential of polyethylene glycol-modified glucagon-like peptide-1s as a type 2 anti-diabetic agent. Regulatory Peptides 2009; 152: 101-107.

[123] Roed SN, Orgaard A, Jorgensen R, De Meyts P. Receptor oligomerization in family B1 of G-protein-coupled receptors: focus on BRET investigations and the link between GPCR oligomerization and binding cooperativity. Front Endocrinol 2012; 3: 62.

[124] Schelshorn D, Joly F, Mutel S, *et al*. Lateral allosterism in the glucagon receptor family: glucagon-like peptide 1 induces G-protein-coupled receptor heteromer formation. Mol Pharmacol 2012; 81: 309-318.

[125] Whitaker GM, Lynn FC, McIntosh CH, Accili EA. Regulation of GIP and GLP1 receptor cell surface expression by N-glycosylation and receptor heteromerization. PLoS ONE 2012; 7: e32675.

[126] Pocai A. Mechanism of action of oxuntomodulin. J Endocrinol 2012; 215: 335-346.

[127] Shekelle PG, Morton SC, Maglione M, *et al*. Pharmacological and surgical treatment of obesity. Evid Rep Technol Assess 2004; 103: 1-6.

[128] Bays H. Phentermine, topiramate and their combination for the treatment of adiposopathy ('sick fat') and metabolic disease. Expert Rev Cardiovasc Ther 2010; 8: 1777-1801.

[129] Topamax. Package insert. Titusville, NJ, USA: Janssen Pharmaceuticals, Inc.; October 2012.

[130] Topamax. Package leaflet. High Wycombe, Bucks: Janssen-Cilag Ltd.; September 2012.

[131] Li Z, Maglione M, Tu W, *et al*. Meta-analysis: pharmacologic treatment of obesity. Ann Intern Med 2005; 142: 532-546.

[132] Silberstein SD, Neto W, Schmitt J, Jacobs D, MIGR-001 Study Group. Topiramate in migraine prevention: results of a large controlled trial. Arch Neurol 2004; 61: 490-495.

[133] Sugrue MF. Pharmacological and ocular hypotensive properties of topical carbonic anhydrase inhibitors. Prog Retin Eye Res 2000; 19: 87-112.

[134] Turenius CI, Htut MM, Prodon DA, *et al*. GABA(A) receptors in the lateral hypothalamus as mediators of satiety and body weight regulation. Brain Res 2009; 1262: 16-24.

[135] Xu Y, O'Brien WG 3rd, Lee CC, Myers MG Jr, Tong Q. Role of GABA release from leptin receptor-expressing neurons in body weight regulation. Endocrinology 2012; 153: 2223-2233.

[136] Khazaal Y, Zullino DF. Topiramate-induced weight loss is possibly due to the blockade of conditioned and automatic processes. Eur J Clin Pharmacol 2007; 63: 891-892.

[137] McElroy SL, Arnold LM, Shapira NA, *et al*. Topiramate in the treatment of binge eating disorder associated with obesity: a randomized, placebo-controlledtrial. Am J Psychiatry 2003; 160: 255-261.

[138] Khazaal Y, Zullino DF. Topiramate in the treatment of compulsive sexual behavior: case report. BMC Psychiatry 2006; 6: 22.

[139] Turenius CI, Htut MM, Prodon DA, *et al*. GABA(A) receptors in the lateral hypothalamus as mediators of satiety and body weight regulation. Brain Res 2009; 1262: 16-24.

[140] Klein S, Burke LE, Bray GA, *et al*. Clinical implications of obesity with specific focus on cardiovascular disease: a statement for professionals from the American Heart Association Council on Nutrition, Physical Activity, and Metabolism: endorsed by the American College of Cardiology Foundation. Circulation 2004; 110: 2952-2967.

[141] Husum H, Van Kammen D, Termeer E, Bolwig G, Mathé A. Topiramate normalizes hippocampal NPY-LI in flinders sensitive line 'depressed' rats and upregulates NPY, galanin, and CRH-LI in the hypothalamus: implications for mood-stabilizing and weight loss-inducing effects. Neuropsychopharmacology 2003; 28: 1292-1299.

[142] Schütt M, Brinkhoff J, Drenckhan M, Lehnert H, Sommer. Weight reducing and metabolic effects of topiramate in patients with migraine--an observational study. Exp Clin Endocrinol Diabetes 2010; 118: 449-452.

[143] Theisen FM, Beyenburg S, Gebhardt S, *et al*. A prospective study of body weight and serum leptin levels in patients treated with topiramate. Clin Neuropharmacol 2008; 31: 226-230.

[144] Tremblay A, Chaput JP, Bérubé-Parent S, *et al*. The effect of topiramate on energy balance in obese men: a 6-month double-blind randomized placebo-controlled study with a 6-month open-label extension. Eur J Clin Pharmacol 2007; 63: 123-134.

[145] Picard F, Deshaies Y, Lalonde J, Samson P, Richard D. Topiramate reduces energy and fat gains in lean (Fa/?) and obese (fa/fa) Zucker rats. Obes Res 2000; 8: 656-663.

[146] Winum JY, Scozzafava A, Montero JL, Supuran CT. Sulfamates and their therapeutic potential. Med Res Rev 2005; 25: 186-228.

[147] Poulsen SA, Wilkinson BL, Innocenti A, Vullo D, Supuran CT. Inhibition of human mitochondrial carbonic anhydrases VA and VB with para-(4-phenyltriazole-1-yl)-benzenesulfonamide derivatives. Bioorg Med Chem Lett 2008; 18: 4624-4627.

[148] Vullo D, Franchi M, Gallori E, Antel J, Scozzafava A, Supuran CT. Carbonic anhydrase inhibitors. Inhibition of mitochondrial isozyme V with aromatic and heterocyclicsulfonamides. J Med Chem 2004; 47: 1272-1279.

[149] Richard D, Ferland J, Lalonde J, *et al*. Influence of topiramate in the regulation of energy balance. Nutrition 2000; 16: 961-966.

[150] Glazer G. Long-term pharmacotherapy of obesity 2000: a review of efficacy and safety. Arch Intern Med 2000; 161: 1814-1824.

[151] Rothman RB, Ayestas MA, Dersch CM, Baumann MH. Aminorex, fenfluramine, and chlorphentermine are serotonin transporter substrates. Implications for primary pulmonary hypertension. Circulation 1999; 100: 869-875.

[152] Rothman RB, Baumann MH, Dersch CM, *et al*. Amphetamine-type central nervous system stimulants release norepinephrine more potently than they release dopamine and serotonin. Synapse 2001; 39: 32-41.

[153] Zolkowska D, Rothman RB, Baumann MH. Amphetamine analogs increase plasma serotonin: implications for cardiac and pulmonary disease. J Pharmacol Exp Ther 2006; 318: 604-610.

[154] Birkenfeld AL, Schroeder C, Boschmann M, *et al*. Paradoxical effect of sibutramine on autonomic cardiovascular regulation. Circulation 2002; 106: 2459-2465.

[155] Eisenhofer G, Saigusa T, Esler MD, Cox HS, Angus JA, Dorward PK. Central sympathoinhibition and peripheral neuronal uptake blockade after desipramine in rabbits. Am J Physiol 1991; 260 (4 Pt 2): R824- R832.

[156] Jordan J, Scholze J, Matiba B, Wirth A, Hauner H, Sharma AM. Influence of sibutramine on blood pressure: evidence from placebo-controlled trials. Int J Obes 2005; 29: 509-516.

[157] Schroeder C, Tank J, Boschmann M, *et al.* Selective norepinephrine reuptake inhibition as a human model of orthostatic intolerance. Circulation 2002; 105: 347-353.

[158] Tank J, Schroeder C, Diedrich A, *et al.* Selective impairment in sympathetic vasomotor control with norepinephrine transporter inhibition. Circulation 2003; 107: 2949-2954.

[159] Esler MD, Wallin G, Dorward PK, *et al.* Effects of desipramine on sympathetic nerve firing and norepinephrine spillover to plasma in humans. Am J Physiol 1991; 260: R817-R823.

[160] Hendricks EJ, Rothman RB, Greenway FL. How physician obesity specialists use drugs to treat obesity. Obesity (Silver Spring) 2009; 17: 1730-1735.

[161] Kang JG, Park CY, Kang JH, Park YW, Park SW. Randomized controlled trial to investigate the effects of a newly developed formulation of phentermine diffuse-controlled release for obesity. Diabetes Obes Metab 2010; 12: 876-882.

[162] Kim KK, Cho HJ, Kang HC, Youn BB, Lee KR. Effects on weight reduction and safety of short-term phentermine administration in Korean obese people. Yonsei Med J 2006; 47: 614-625.

[163] Munro JF, MacCuish AC, Wilson EM, Duncan LJ. Comparison of continuous and intermittent anorectic therapy in obesity. BMJ 1968; 1: 352-354.

[164] Hendricks EJ, Greenway FL, Westman EC, Gupta AK. Blood pressure and heart rate effects, weight loss and maintenance during long-term phentermine pharmacotherapy for obesity. Obesity (Silver Spring) 2011; 19: 2351-2360.

[165] Hendricks EJ, Srisurapanont M, Schmidt SL, *et al.* Addiction potential of phentermine prescribed during long-term treatment of obesity. Int J Obes (Lond) 2014; 38: 292-298.

[166] Kennett GA, Clifton PG. New approaches to the pharmacological treatment of obesity: can they break through the efficacy barrier? Pharmacol Biochem Behav 2010; 97: 63-83.

[167] Gadde KM, Allison DB, Ryan DH, *et al.* Effects of low dose, controlled release, phentermine plus topiramate combination on weight and associated comorbidities in overweight and obese adults (CONQUER): a randomised, placebo-controlled, phase 3 trial. Lancet 2011; 377: 1341-1352.

[168] Garvey WT, Ryan DH, Look M, *et al.* Two-year sustained weight loss and metabolic benefits with controlled release phentermine/topiramate in obese and overweight adults (SEQUEL): a randomized, placebo-controlled, phase 3 extension study. Am J Clin Nutr 2012; 95: 297-308.

[169] Center for Disease Control. Cardiac valvulopathy associated with exposure to fenfluramine or dexfenfluramine: U.S. Department of Health and Human Services Interim Public Health Recommendations. MMWR Morb Mortal Wkly Rep 1997; 46: 1061-1066.

[170] Gadde KM, Allison DB, Ryan DH, *et al.* Effects of low-dose, controlled-release, phentermine plus topiramate combination on weight and associated comorbidities in overweight and obese adults (CONQUER): a randomised, placebocontrolled, phase 3 trial. Lancet 2011; 377: 1341-1352.

[171] Garvey WT, Ryan DH, Look M, *et al.* Two-year sustained weight loss and metabolic benefits with controlled-release phentermine/topiramate in obese and overweight adults (SEQUEL): a randomized, placebo-controlled, phase 3 extension study. Am J Clin Nutr 2012; 95: 297-308.

[172] Jordan J, Troupin BT. Cardiovascular effects and weight loss in three 1-year, double-blind, placebo-controlled clinical trials with extended release phentermine/topiramate in obese patients. Presented at the 22nd Annual Congress of the European Society of Hypertension (ESH). London, UK; April 2012.

[173] Jordan J, Astrup A, Day W. Cardiovascular safety of phentermine alone and in combination with topiramate. Presented at the 19th Annual European Congress on Obesity (ECO). Lyon, France; May 2012.

[174] Bays H. Phentermine, topiramate and their combination for the treatment of adiposopathy ('sick fat') and metabolic disease. Expert Rev Cardiovasc Ther 2010; 8: 1777-1801.

[175] Brownell KD. The LEARN program for weight management, 10th ed American Health Publishing Company; 2004.

[176] Wyatt HR. Update for treatment strategies of obesity. J Clin Endocrinol Metabol 2013 [Epub ahead of print February 26, 2013 as doi: 10.1210/jc.2012-3115].

[177] Aronne LJ, Wadden TA, Peterson C, Winslow D, Odeh S, Gadde KM. Evaluation of phentermine and topiramate versus phentermine/topiramate extended-release in obese adults. Obesity (Silver Spring) 2013 Nov;21(11): 2163-71. doi: 10.1002/oby.20584. Epub 2013 Oct 17.

[178] U.S. Food and Drug Administration. Briefing Information for the February 22, 2012 Meeting of the Endocrinologic and Metabolic Drugs Advisory Committee (WWW document); 2012. http: //www.fda.gov/AdvisoryCommittees/CommitteesMeetingMaterials/Drugs/Endocrinologicand MetabolicDrugsAdvisoryCommittee/ucm292314.htm.

[179] Davidson M, Bowden CH, Day WW. Weight loss and cardiovascular risk reduction over 2 years with controlled-release phentermine- topiramate. Presented at the 60th Annual Scientific Session and Expo of the American College of Cardiology (ACC). New Orleans; April 2011.

[180] Vivus Inc. [Accessed June 28, 2013] Qsymia (phentermine and topiramate extended-release) capsules, for oral use. 4/16/2013; http: //www.accessdata.fda.gov/drugsatfda_docs/label/2013/022580s004lbl.pdf.

[181] Garvey WT, Ryan DH, Look M, *et al*. Two-year sustained weight loss and metabolic benefits with controlled-release phentermine/topiramate in obese and overweight adults (SEQUEL): a randomized, placebo-controlled, phase 3 extension study. Am J Clin Nutr 2012; 95(2): 297-308.

[182] Allison DB, Gadde KM, Garvey WT, *et al*. Controlled-release phentermine/topiramate in severely obese adults: a randomized controlled trial (EQUIP). Obesity (Silver Spring) 2012; 20(2): 330-342.

[183] Margulis AV, Mitchell AA, Gilboa SM, *et al*. Use of topiramate in pregnancy and risk of oral clefts. Am J Obst Gynecol 2012; 207(5): 405, e401-40.

[184] Vivus Inc. NDA 22580: QSYMIA (phentermine and topiramate extended-release) Capsules. Risk evaluation and mitigation strategy (REMS); Reference ID: 3294731. 4/2013; http: //www.fda.gov/downloads/Drugs/DrugSafety/PostmarketDrugSafetyInformationforPatientsan dProviders/UCM312598.pdf. [Accessed 7/3/2013]

[185] Tran, PT. [Accessed 7/3/2013] Summary Minutes of the Endocrinologic and Metabolic Drugs Advisory Committee Meeting. Feb. 2012 http: //www.fda.gov/downloads/AdvisoryCommittees/CommitteesMeetingMaterials/Drugs/Endocri nologicandMetabolicDrugsAdvisoryCommittee/UCM304401.pdf.

[186] Colman E, Golden J, Roberts M, Egan A, Weaver J, Rosebraugh C. The FDA's assessment of two drugs for chronic weight management. N Engl J Med 2012; 367(17): 1577-1579.

[187] Gadde KM, Allison DB, Ryan DH, *et al*. Effects of low-dose, controlled-release, phentermine plus topiramate combination on weight and associated comorbidities in overweight and obese adults (CONQUER): a randomised, placebo-controlled, phase 3 trial. Lancet 2011; 377(9774): 1341-1352.

[188] Fleming JW, McClendon KS, Riche DM. New obesity agents: lorcaserin and phentermine/topiramate. Ann Pharmacother 2013; 47: 1007-1116.

[189] Qsymia [package insert]. VIVUS, Inc., Mountain View, CA.

[190] Rapport MM, Green AA, Page IH. Crystalline serotonin. Science 1948; 108(2804): 329-330.

[191] Wooley DW, Shaw E. A biochemical and pharmacological suggestion about certain mental disorders. Proc Natl Acad Sci USA 1954; 40(4): 228-231.

[192] Gaddum JH, Picarelli ZP. Two kinds of tryptamine receptors. Br J Pharmacol Chemother 1957; 12(3): 323-328.

[193] Thomsen WJ, Grottick AJ, Menzaghi F, *et al.* Lorcaserin, a novel selective human 5-hydroxytryptamine2C agonist: *in vitro* and *in vivo* pharmacological characterization. J Pharmacol Exp Ther 2008; 325(2): 577-587.

[194] Roth BL, Willins DL, Kristiansen K, Kroeze WK. 5-Hydroxytryptamine2-family receptors (5-hydroxytryptamine2A, 5-hydroxytryptamine2B, 5-hydroxytryptamine2C): where structure meets function. Pharmacol Ther 1998; 79(3): 231-257.

[195] Roth BL, Nakaki T, Chuang DM, Costa E. Aortic recognition sites for serotonin (5HT) are coupled to phospholipase C and modulate phosphatidylinositol turnover. Neuropharmacology 1984; 23(10): 1223-1225.

[196] Conn PJ, Sanders-Bush E. Selective 5-HT2 antagonists inhibit serotonin-stimulated phosphatidylinositol metabolism in cerebral cortex. Neuropharmacology 1984; 23(8): 993-996.

[197] Roth BL, Nakaki T, Chuang DM, Costa E. 5-Hydroxytryptamine2 receptors coupled to phospholipase C in rat aorta: modulation of phosphoinositide turnover by phorbol ester. J Pharmacol Exp Ther 1986; 238(2): 480-485.

[198] Roth BL, Ciaranello RD, Meltzer HY. Binding of typical and atypical antipsychotic agents to transiently expressed 5-HT1C receptors. J Pharmacol Exp Ther 1992; 260(3): 1361-1365.

[199] Roth BL, Sheffler DJ, Kroeze WK. Magic shotguns versus magic bullets: selectively non-selective drugs for mood disorders and schizophrenia. Nat Rev Drug Discov 2004; 3(4): 353-359.

[200] Roth B, Willins D, Kristiansen K, Kroeze W. Activation is hallucinogenic and antagonism is therapeutic: role of 5-HT2A receptors in atypical antipsychotic drug actions. The Neuroscientist 1999; 5: 254-262.

[201] Roth BL. Drugs and valvular heart disease. N Engl J Med 2007; 356(1): 6-9.

[202] Huang XP, Selola V, Yaday PN, *et al.* Parallel functional activity profiling reveals valvulopathogens are potent 5-hydroxytryptamine (2B) receptor agonists: implications for drug safety assessment. Mol Pharmacol 2009; 76(4): 710-722.

[203] Wacker D, Wang C, Katritch V, *et al.* Structural features for functional selectivity at serotonin receptors. Science 2013; 340(6132): 615-619.

[204] Blundell JE. Serotonin and appetite. Neuropharmacology 1984; 23(12B): 1537-1551.

[205] Tecott LH, Sun LM, Asana SF, *et al.* Eating disorder and epilepsy in mice lacking 5-HT2c serotonin receptors. Nature 1995; 374(6522): 542-546.

[206] Vickers SP, Clifton PG, Dourish CT, Tecott LH. Reduced satiating effect of d-fenfluramine in serotonin 5-HT(2C) receptor mutant mice. Psychopharmacology (Berl) 1999; 143(3): 309-314.

[207] Weintraub M, Hasday JD, Mushlin AI, Lockwood DH. A double-blind clinical trial in weight control. Use of fenfluramine and phentermine alone and in combination. Arch Intern Med 1984; 144(6): 1143-1148.

[208] Connolly HM, Grary JL, McGoon MD, *et al.* Valvular heart disease associated with fenfluramine-phentermine. N Engl J Med 1997; 337(9): 581-588.

[209] Brenot F, Herve P, Petitpretz P, Parent F, Duroux P, Simonneau G. Primary pulmonary hypertension and fenfluramine use. Br Heart J 1993; 70(6): 537-541.

[210] Rothman RB, Baumann MH, Savage JE, *et al*. Evidence for possible involvement of 5-HT(2B) receptors in the cardiac valvulopathy associated with fenfluramine and other serotonergic medications. Circulation. 2000; 102(23): 2836-2841.

[211] Launay JM, Herve P, Peoch K, *et al*. Function of the serotonin 5-hydroxytryptamine 2B receptor in pulmonary hypertension. Nat Med 2002; 8(10): 1129-1135.

[212] Fitzgerald LW, Burn TC, Brown BS, *et al*. Possible role of valvular serotonin 5-HT(2B) receptors in the cardiopathy associated with fenfluramine. Mol Pharmacol 2000; 57(1): 75-81.

[213] Giradet C, Butler AA. Neural melanocortin receptors in obesity and related metabolic disorders. Biochim Biophys Acta 2014; 1842(3): 482-494.

[214] Manning S, Pucci A, Finer N. Pharmacotherapy for obesity: novel agents and paradigms. Ther Adv Chronic Dis 2014; 5(3): 135-148.

[215] Martin CK, Redman LM, Zhang J, *et al*. Lorcaserin, a 5-HT(2C) receptor agonist, reduces body weight by decreasing energy intake without influencing energy expenditure. J Clin Endocrinol Metab 2011; 96(3): 837-845.

[216] Lam DD, Przydzial MJ, Ridley SH, *et al*. Serotonin 5-HT2C receptor agonist promotes hypophagia *via* downstream activation of melanocortin 4 receptors. Endocrinology 2008; 149: 1323-1328.

[217] Nichols DE. Hallicinogens. Pharmacol Ther 2004; 101: 131-181.

[218] Fidler MC, Sanchez M, Raether B, *et al*. A one-year randomized trial of lorcaserin for weight loss in obese and overweight adults: the BLOSSOM trial. J Clin Endocrinol Metab 2011; 96(10): 3067-3077.

[219] Smith SR, Weissman NJ, Anderson CM, *et al*. Multicenter, placebo-controlled trial of lorcaserin for weight management. N Engl J Med 2010; 363(3): 245-256.

[220] Belviq (lorcaserin hydrochloride) package insert. Zofingen, Switzerland: Arena Pharmaceuticals GmbH; 2012.

[221] O'Neil PM, Smith SR, Weissman SJ, *et al*. Randomized placebo-controlled clinical trial of lorcaserin for weight loss in type 2 diabetes mellitus: the BLOOM-DM study. Obesity (Silver Spring) 2012; 20(7): 1426-1436.

[222] Chan E, He Y, Chui C, Wong A, Lau W, Wong I. Efficacy and safety of lorcaserin in obese adults: a meta-analysis of 1-year randomized controlled trials and narrative review on short term randomized controlled trials. Obes Rev 2013; 14: 383-392.

[223] Kublaoui BM, Gemelli T, Tolson KP, *et al*. Oxytocin deficiency mediates hyperphagic obesity of Sim1 haploinsufficient mice. Mol Endocrinol 2008; 22: 1723-1734.

[224] Kublaoui BM, Holder JL, Jn, Gemelli T, Zinn AR. Sim1 haploinsufficiency impairs melanocortin-mediated anorexia and activation of paraventricular nucleus neurons. Mol Endocrinol 2006; 20: 2483-2492.

[225] Tolson KP, Gemelli T, Gautron L, *et al*. Postnatal Sim1 deficiency causes hyperphagic obesity and reduced Mc4r and oxytocin expression. J Neurosci 2010; 30: 3803-3812.

[226] Swaab DF, Purba JS, Hofman MA. Alterations in the hypothalamic paraventricular nucleus and its oxytocin neurons (putative satiety cells) in Prader-Willi syndrome: a study of five cases. J Clin Endocrinol Metab 1995; 80: 573-579.

[227] Baskin DG, Kim F, Gelling RW, *et al*. A new oxytocin-saporin cytotoxin for lesioning oxytocin-receptive neurons in the rat hindbrain. Endocrinology 2010; 151: 4207-4213.

[228] Zhang G, Bain H, Zhang H, *et al*. Neuropeptide exocytosis involving synaptotagmin-4 and oxytocin in hypothalamic programming of body weight and energy balance. Neuron 2011; 69: 523-535.

[229]   Deblon N, Veyrat-Durebex C, Bourqoin M, *et al.* Mechanisms of the anti-obesity effects of oxytocin in diet-induced obese rats. PLoS One 2011; 6: e25565.

[230]   Zhang G, Cai D. Circadian intervention of obesity development via resting-stage feeding manipulation or oxytocin treatment. Am J Physiol Endocrinol Metab 2011; 301: E1004-E1012.

[231]   Morton GJ, Thatcher BS, Reidelberger RD, *et al.* Peripheral oxytocin suppresses food intake and causes weight loss in diet-induced obese rats. Am J Physiol Endocrinol Metab 2012; 302: E134-E144.

[232]   Adan RA. Mechanisms underlying current and future anti-obesity drugs. Trends Neurosci 2013; 36: 133-140.

[233]   Kosfeld M, Heinrichs M, Zak PJ, *et al.* Oxytocin increases trust in humans. Nature 2005; 435: 673-676.

[234]   Feifel D, Macdonald K, Nguyen A, *et al.* Adjunctive intranasal oxytocin reduces symptoms in schizophrenia patients. Biol Psychiatry 2010; 68: 678-680.

[235]   Zhang H, Wu C, Chen Q, *et al.* Treatment of obesity and diabetes using oxytocin or analogs in patients and mouse models. PLoS One 2013; 8(5): e61477.

[236]   Han L, Zhou R, Niu J, *et al.* SIRT1 is regulated by a PPAR{gamma}-SIRT1 negative feedback loop associated with senescence. Nucleic Acids Res 2010; 38: 7458-7471.

[237]   Wang C, Powell MJ, Popov VM, Pestell RG. Acetylation in nu403. Modulating PPARG to control inflammation associated with diabetes http: //e-dmj.org. Mol Endocrinol 2008; 22: 539-545.

[238]   Choi JH, Banks AS, Estall JL, *et al.* Anti-diabetic drugs inhibit obesity-linked phosphorylation of PPAR gamma by Cdk5. Nature 2010; 466: 451-456.

[239]   Kamenecka TM, Busby SA, Kumar N, *et al.* Potent anti-diabetic actions of a novel non-agonist PPAR gamma ligand that blocks Cdk5-mediated phosphorylation. Available from: http: //www.ncbi.nlm.nih.gov/books/NBK143191/pdf/ml244.pdf (updated 2013 Mar 7).

[240]   Choi JH, Banks AS, Kamenecka TM, *et al.* Antidiabetic actions of a non-agonist PPARgamma ligand blocking Cdk5-mediated phosphorylation. Nature 2011; 477: 477-481.

[241]   Adam, J.A. Menheere P, Van Dielen FM, *et al.* Decreased plasma orexin-A levels in obese individuals. Int J Obes Relat Metab Disord 2002; 26: 274-276.

[242]   Baranowska, BA, Baranowska B, Wolinska-Witort E, *et al.* Plasma orexin A, orexin B, leptin, neuropeptide Y (NPY) and insulin in obese women. Neuro Endocrinol Lett 2005; 26: 293-296.

[243]   Bronsky J, Nedvidkova J, Zamrazilova H, *et al.* Dynamic changes of orexin A and leptin in obese children during body weight reduction. Physiol Res 2007; 56: 89-96.

[244]   Smart D, Haynes AC, Williams G, Arch JR. Orexins and the treatment of obesity. Eur J Pharmacol 2002; 440: 199-212.

[245]   Sellayah D, Sikder D. Feeding the heat on brown fat. Ann N Y Acad Sci 2013; 1302: 11-23.

[246]   Astrup A, Madsbad S, Breum L, *et al.* Effect of tesofensine on body weight loss, body composition, and quality of life in obese patients: a randomized double blind, placebo-controlled trial. Lancet 2008; 372: 1906-1913.

[247]   Astrup A. Drug management of obesity - efficacy versus safety. N Engl J Med 2010; 363: 288-290.

[248]   George M, Rajaram M, Shanmugam E. New and emerging drug molecules against obesity. J Cardiovasc Pharmacol Ther 2014; 19(1): 65-76. doi: 10.1177/1074248413501017. Epub 2013 Sep 24.

[249]   Kim YM, An JJ, Jin YJ, *et al.* Assessment of the anti-obesity effects of the TNP-470 analog, CKD-732. J Mol Endocrinol 2007; 38(4): 455-465.

[250] Datta B, Majumdar A, Datta R, Balusu R. Treatment of cells with the angiogenic inhibitor fumagillin results in increased stability of eukaryotic initiation factor 2-associated glycoprotein, p67, and reduced phosphorylation of extracellular signal-regulated kinases. Biochemistry 2004; 43(46): 14821-14831.

[251] Raghow R, Yellaturu C, Deng X, Park EA, Elam MB. SREBPs: the crossroads of physiological and pathological lipid homeostasis. Trends Endocrinol Metab 2008; 19(2): 65-73.

[252] Kotzka J, Knebel B, Avci H, *et al.* Phosphorylation of sterol regulatory element-binding protein (SREBP)-1a links growth hormone action to lipid metabolism in hepatocytes. Atherosclerosis 2010; 213(1): 156-165.

[253] Zafgen [website on the Internet]. Press releases. Cambridge: Zafgen; 2013. Available from: http: //zafgen.com/zafgen/newsroom/press-releases. Accessed October 22, 2013.

[254] Cao Y. Angiogenesis modulates adipogenesis and obesity. J Clin Invest 2007; 117(9): 2362-2368.

[255] Bukowiecki L, Lupien J, Follea N, *et al.* Mechanism of enhanced lipolysis in adipose tissue of exercise-trained rats. Am J Physiol 1980; 239(6): E422-E429.

[256] Gupta RK, Mepani RJ, Kleiner S, *et al.* Zfp423 expression identifies committed preadipocytes and localizes to adipose endothelial and perivascular cells. Cell Metab 2012; 15(2): 230-239.

[257] Cao Y. Endogenous angiogenesis inhibitors and their therapeutic implications. Int J Biochem Cell Biol 2001; 33(4): 357-369.

[258] Sun K, Wernstedt Asterholm I, Kusminski CM, *et al.* Dichotomous effects of VEGF-A on adipose tissue dysfunction. Proc Natl Acad Sci USA 2012; 109(15): 5874-5879.

[259] Kuo LE, Kitlinska JB, Tilan JU, *et al.* Neuropeptide Y acts directly in the periphery on fat tissue and mediates stress-induced obesity and metabolic syndrome. Nat Med 2007; 13(7): 803-811.

[260] Tam J, Duda DG, Perentes JY, *et al.* Blockade of VEGFR2 and not VEGFR1 can limit diet-induced fat tissue expansion: role of local versus bone marrow-derived endothelial cells. PLoS One 2009; 4(3): e4974.

[261] Cao R, Brakenhielm E, Wahlestedt C, Thyberg J, Cao Y. Leptin induces vascular permeability and synergistically stimulates angiogenesis with FGF-2 and VEGF. Proc Natl Acad Sci USA 2001; 98(11): 6390-6395.

[262] Joharapurkar AA, Dhanesha NA, Jain MR. Inhibition of the methionine aminopeptidase 2 enzyme for the treatment of obesity. Diabetes, Metabolic Syndrome and Obesity: Targets and Therapy 2014; 7: 73-84.

[263] Rupnick MA, Panigrahy D, Zhang CY, *et al.* Adipose tissue mass can be regulated through the vasculature. Proc Natl Acad Sci USA 2002; 99(16): 10730-10735.

[264] Kopelman P, Bryson A, Hickling R, *et al.* Cetilistat (ATL-962), a novel lipase inhibitor: a 12-week randomized, placebo-controlled study of weight reduction in obese patients. Int J Obes 2007; 31: 494-499.

[265] Kopelman P, Groot GH, Rissanen A, *et al.* Weight loss, HbA1c reduction, and tolerability of cetilistat in a randomized, placebo-controlled phase 2 trial in obese diabetics: comparison with orlistat (Xenical). Obesity 2010; 18: 108-115.

[266] Greenway FL, Dunayevich E, Tollefson G, *et al.* Comparison of combined bupropion and naltrexone therapy for obesity with monotherapy and placebo. J Clin Endocrinol Metab 2009; 94: 4898-4906.

[267] Greenway FL, Fujioka K, Plodkowski RA, *et al*. Effect of naltrexone plus bupropion on weight loss in overweight and obese adults (COR-I): a multicentre, randomised, double-blind, placebo-controlled, phase 3 trial. Lancet 2010; 376: 595-605.

[268] Wadden TA, Foreyt JP, Foster GD, *et al*. Weight loss with naltrexone SR/bupropion SR combination therapy as an adjunct to behavior modification: the COR-BMOD trial. Obesity (Silver Spring) 2011; 19: 110-120.

[269] Gadde KM, Yonish GM, Foust MS, Wagner HR. Combination therapy of zonisamide and bupropion for weight reduction in obese women: a preliminary, randomized, open-label study. J Clin Psych 2007; 68: 1226-1229.

[270] Ioannides Demos LL, Piccenna L, McNeil JJ. Pharmacotherapies for Obesity: Past, Current, and Future Therapies. J Obes 2011 doi: 10.1155/2011/179674.

[271] Lucio Cabrerizo García1, Ana Ramos-Leví1, Carmen Moreno Lopera, Miguel A. Rubio Herrera. Update on pharmacology of obesity: Benefits and risks. Nutr Hosp 2013; 28(Suppl.5): 121-127.

Subject Index

A

Abdominal obesity 120, 155, 171
Abnormal ALT 158
Absorption of carbohydrates 4, 55
Acarbose-metformin 168
Activating GK mutations 121
Active AdipoR 27
Active comparators 162, 163
Active GLP-1 7, 165, 166
Acute 126, 173, 174, 197
   myocardial infarction 126
   pancreatitis 173, 174, 197
Alogliptin 170
   addition of   170
Adipocytes 26, 27, 30, 32, 83, 85, 86, 90, 91, 94, 96, 97, 199
Adipogenesis 132, 190, 211, 214
Adipokines 30, 84, 89, 103, 211
Adiponectin 9, 27, 83, 84, 85, 86, 87, 89, 90, 94, 129, 210
Adiponectin expression 84, 85, 86, 89
Adiponectin mimetics 3, 26, 27
Adiponectin receptor 27, 33, 89, 90, 94
Adipose angiogenesis 214
Adipose-derived precursor cells 190
Adipose tissue 26, 28, 30, 50, 52, 53, 54, 55, 88, 89, 90, 97, 129, 130, 131
Adjusted hazard ratio 21
Adverse cardiovascular events 162, 171, 173
Albiglutide 54, 105, 106, 107, 108, 161, 162, 197
Aleglitazar 134, 135
Alendronate 93, 96
Alogliptin 7, 54, 111, 112, 150, 164, 169, 170, 171, 173, 175
Alpha2A-adrenergic 120, 121
   receptor antagonists 120, 121
   receptors 120
Alpha-glucosidase inhibitors 55, 61
Ameliorate diabetes 27
Ameliorate insulin receptor signaling 211
Amelioration 8, 9, 203
Amino acid sequence 8, 106
Amphetamine 24, 203, 204, 217
AMPK 52, 53, 129
   activation 129
   levels 52, 53
   mediated glucose uptake in muscle cells 129
Angiogenesis 26, 28, 190, 214
Anorectic actions 204

Antidiabetic 25, 50, 51, 60, 61, 120, 124, 127, 128, 129, 140, 150, 165, 169, 210
   agents 51, 60, 61, 124, 127, 128, 169
   drugs 50, 120, 129, 150
   drugs rosiglitazone 25
   effects 165
   properties 210
   therapies 140
Anti-obesity 3, 12, 23, 104, 135, 189, 214, 216, 217
   agents 214, 216
   drugs 3, 12, 23, 135, 189, 216, 217
   drug QNEXA 104
   regimen 205, 214
Antiresorptive drugs 96
Antisense oligonucleotides 114, 123
Apoptosis 125, 152
Appetite 3, 14, 24, 138, 152, 154, 196, 199, 204, 205, 208
Asparagine 11
Atf4 overexpression in osteoblasts 91
Atherosclerosis 27, 28, 32, 91, 103

B

Bariatric procedures 138, 139
Bariatric surgeries 138, 139
   Basal 12, 20, 56, 159, 169
   insulin 12, 20, 56, 159
   insulin therapy 169
Baseline 12, 66 69,, 155, 168
   HbA1C 12, 66, 155, 168
   insulin 69
β-cell 70, 102, 103, 112, 119, 124, 125, 128, 137, 150, 152, 153, 154, 163, 166, 167, 168, 176
   failure 103, 124
   function 70, 102, 103, 112, 119, 125, 128, 150, 154, 166, 167, 168, 176
   proliferation 152, 153, 163
   regeneration 137
Beloranib 189, 190, 212, 213, 214, 217
Betatrophin 3, 26, 29, 30
β-glucosidase enzymes 114
Biguanides 3, 9, 61
Bile acid synthesis 31
6-bisphosphatase 126, 127
Blood glucose levels 4, 21, 55, 60, 91, 93, 102, 104, 133, 151
Blood pressure (BP) 76, 120, 131, 132, 136, 156, 157, 201, 203, 213, 217
Body mass index (BMI) 13, 14, 23, 24, 34, 50, 51, 136, 138, 139, 202, 206, 213, 215, 217